THE AMERICAN IDEOLOGY
OF NATIONAL SCIENCE,
1919-1930

RONALD C. TOBEY

THE AMERICAN IDEOLOGY OF NATIONAL SCIENCE, 1919-1930

UNIVERSITY OF PITTSBURGH PRESS

Library of Congress Catalog Card Number 70-151507
ISBN 0-8229-3227-X
Copyright © 1971, University of Pittsburgh Press
All rights reserved
Henry M. Snyder & Co., Inc., London
Manufactured in the United States of America

710 424

TO SUE

CONTENTS

ACKNOWLEDGMENTS

The author appreciates the aid of a number of persons in the preparation of this book. David Davis encouraged the study in all its stages. He made valuable criticisms of chapters 1 and 7 which brought the refocusing of their themes. Daniel Kevles made the author's research at the California Institute of Technology more profitable and enjoyable than it otherwise would have been. He was responsible for bringing the Hale correspondence in the Huntington Library to the author's attention. Nathan Reingold carefully read the entire manuscript, helping to eliminate some unclear statements and errors. T. Owen Carroll read chapter 4 on the Einstein controversy. His criticism was responsible for correcting the author's misinterpretation of a mathematical aspect of the theories of relativity. The author is also indebted to Laurans Laudan, Jonathan Levine, Robert Doherty, Carter Goodrich, Samuel Hays, and Roy Lubove.

The librarians and archivists at the following institutions are thanked for their labors on behalf of this study: the California Institute of Technology Archives, the Cornell University Library and Archives, the Huntington Library, and the Library of Congress Manuscript Division.

While undertaking the initial stages of this work, the author was supported by a Woodrow Wilson Dissertation Fellowship. The Wilson fellowship particularly facilitated travel for research.

The author owes a special gratitude for the contributions of his wife to this study.

INTRODUCTION

The division between the scientific and nonscientific cultures in America involves both the specialization of knowledge and the fragmentation of values. Social cohesion is threatened not only by the inability of the nonscientists to understand what the scientists are doing, but also by the inability of the one to share or to sympathize with the values of the other. Although the division was well advanced in the nineteenth century, the years from 1919 to 1930 were critical for its modern form because of two developments. First, leading American scientists were prompted by their experiences in World War I to reorganize science on a new basis that required a correspondence between their professional values and broader political and cultural values. Second, the lay public witnessed a controversy over Einstein's theories of relativity. The theories made profound and controversial changes in the character of physical science which, in turn, made it difficult to relate scientific values to the cultural values and everyday life of the nonscientist.

The origin of the contemporary form of the two-cultures problem is analyzed in the terms of the institutional and intellectual situation that motivated leading scientists in the 1920s to seek a public consensus on an ideology of national science. *National science* has two related meanings. It refers to the centralized administration of nongovernmental scientific activity by a private agency of the kind that developed during the First World War for war-related research. It refers also to the articulated, explicit relevance of the values of professional science to the values of nonscientists, whether the latter be the officially sponsored ideals of the federal government in the First World War or the economic and political values which the scientists thought the general public held in the 1920s. This study deals primarily with national science in the second meaning and discusses national science in the first meaning to the extent necessary to understand the motivations of the leading scientists. The purpose of the study does not include narrating, beyond the boundaries of the basic cultural situation, the professional history of the sciences and the history of scientific institutions, or describing the scientists' world view.

In the progressive period, professional, nongovernmental science was culturally isolated and decentralized. Scientific activity was located in a variety of philanthropically supported institutions and universities which had few ties to each other or to the government. The scientists themselves were organized into working societies in their special fields, and most of the activity of these societies was conducted at the local or regional level. At the same time, the professionalization and specialization of science which had occurred in the nineteenth century removed the daily work of the scientists from public concern.

The First World War altered this situation in important ways. The National Research Council, founded during the preparedness movement, centralized the administration of war-related research. Many nongovernmental scientists were recruited to work on defense problems such as submarine detection. The values of professional scientific research were related to the national goals of the war effort, thus providing a rhetorical expression of cultural unity deeply satisfying to scientists. For a few scientists like George Ellery Hale, the war effort reaffirmed a long-standing belief that the progress of American science in all fields would benefit from central administration by a private agency like the National Research Council. In other scientists like Robert A. Millikan, the war generated a new commitment to coordinated research. While there was disagreement among leading scientists over the degree of centralization and the precise role of central institutions, there was agreement that American science after the war would best be promoted by the continued work of the Research Council, by additional funds gathered and dispersed by a central agency representing all fields, and by the continued relevance of professional values to broader cultural values.

The end of the war, the political struggle over the peace treaty, and the growing disillusionment with the war's official ideals, however, brought an end to the obvious relevance of the scientist to national goals. There was no longer a great enterprise like the war effort conducted by the government in which the nongovernmental scientists had a role and which would justify the new organization of science. With the expectation of restoring the lost correspondence between their values and broad cultural values and of obtaining new sources of financial support, leading scientists endeavored to convince the public that the scientific method was the ultimate guarantee of the existence of the values of prewar progressivism—individualism, political and economic democracy, and progress.

Efforts to build the public consensus were undermined by the agitation over Einstein's theories of relativity. In his popularization of science, for instance, Edwin E. Slosson tried to win the sympathies of the layman for the ideology of national science. But the theories of relativity confused and alienated the nonscientists. Robert Millikan argued that the scientific method revealed that progress was a principle of the universe. The theories of relativity denied science the power of revelation.

The attempt to relate the values of pure science to the values of progressivism was thwarted by the disintegration of that movement in the 1920s. National science was not sufficiently strong an ideology to provide coherence for the declining reform philosophy. At the same time, a campaign to raise money for national scientific research forced the scientists toward conservatism as they sought to accommodate themselves to the values of big-business donors.

At the moment when the division between the two cultures became intolerable to the scientists, the changes in the character of modern science and modern society made it impossible for the two cultures to be unified. The scientists were only partially successful in reorganizing science on a national basis; furthermore, this reorganization did not rest on a shared vision of progressive democracy.

THE AMERICAN IDEOLOGY
OF NATIONAL SCIENCE,
1919-1930

(1) THE PARADOX OF PROGRESSIVE SCIENCE

In the seventeen years preceding America's entry into the First World War, popular interest in science and professional popularization remained at a low level. The great age of popular science generated by the controversy over evolutionary theory had passed. After 1900, no Louis Agassiz, John Fiske, or T. H. Huxley toured the nation creating enthusiasm for science. The eruption of public anxiety over physical science in the 1920s was still to come. In the progressive era, the daily advance of research was of peripheral concern, only occasionally capturing public attention in the form of a quasi science like scientific management.

Nevertheless, it was paradoxical that the waning of popular interest in science and the decline of popularization coincided with the triumph of the reform philosophy of progressivism which assigned a central role to science. According to the major statements of this philosophy, such as Herbert Croly's *The Promise of American Life* (1909) and Walter Lippmann's *Drift and Mastery* (1914), science provided a model for the reform of society and, as well, techniques making democracy possible. Scientific rationalism had a key role in providing coherence for otherwise disparate reform activities. Both the impact of the First World War on the scientists and the impact of the revolution in physics on American culture in the 1920s were conditioned by this peculiar prewar situation—the disinterest of the public in science and the cultural isolation of basic, natural scientists, in a society which self-consciously made science one of its primary values.

(3)

The Decline of Popular Science

In the progressive era, scientists perceived an increasing, if vague, public esteem and felt honor in their profession. The *Nation* corroborated this perception by noting that "there was never so general an impulse to pay honor to science, and do homage to the scientist, as exists at the present time."[1] W. G. Farlow, president of the American Association for the Advancement of Science, thought that a favorable change in public attitudes toward scientists had recently occurred.[2]

Several developments could have initiated this change. The Nobel prizes, first awarded in 1901, stimulated an increase in the prestige of scientists. Farlow suggested that the public esteem for scientists had primarily been raised by their association with the great captains of American industry.[3] Certainly the foundation of the Rockefeller Institute for Medical Research and the Carnegie Institution enhanced the prestige of the research which these were to conduct. The incorporation of the Rockefeller Institute in June 1901 was fully discussed by the New York newspapers, which reported that the institute would help remove America's deficiency in original research and would give the most hope for medical discoveries.[4] Because of the reputation of European medical research institutes, such as the Pasteur Institute and Koch Institute, public esteem for the Rockefeller Institute was assured.

The Carnegie Institution elicited similar editorial praise. Resulting from the merger of the Washington Academy of Sciences's plan to stimulate research, the George Washington Memorial Association's desire for a memorial to the first president, and Andrew Carnegie's wish to found an educational institution, the ten-million-dollar Carnegie Institution bequest was

1. "Popular Appreciation of Scientists," *Nation* 74 (January 16, 1902): 47.
2. W. G. Farlow, "The Popular Conception of the Scientific Man at the Present Day," *Science*, n.s. 23 (January 5, 1906): 2–3.
3. Ibid., p. 3.
4. The director of the institute who briefed the newspapers was T. Mitchell Prudden, a pathologist at Columbia University's College of Physicians. George W. Corner, *A History of the Rockefeller Institute, 1901–1953: Origins and Growth* (New York: Rockefeller Institute Press, 1964), p. 38.

announced in December 1901. The trust deed explictly stated that the aim of the institution was to promote original research in any field without regard to the eventual utility of the investigations. Initial hostility to Carnegie's gift came from the misplaced fear that it might endow a national university, though this was not its purpose. Aside from this, the Carnegie Institution received accolades similar to those for the Rockefeller Institute. The *Independent*, the *Nation*, the New York *Daily Tribune*, and the *New York Times* considered that the institution would greatly promote and enhance research in America.[5]

The Louisiana Purchase Exposition of 1904 in St. Louis similarly honored science. Most accounts of the fair considered the international Congress of Arts and Sciences its most important event. The rare dissent from the praise for the congress was made by Edwin Slosson who thought the papers delivered by the internationally famous scientists spoke for the disunity of the sciences, rather than their unity, the theme of the congress. For American scientists, the congress was an opportunity to host the great European scientists and to receive personal honors in return, as when Ludwig Boltzmann, the great Viennese physicist, visited the unknown physicist Robert A. Millikan at his University of Chicago laboratory.[6]

Public honors and vague esteem notwithstanding, both scientists and the intellectual journals were mildly uneasy about public ignorance of current developments in science and the common

5. C. D. Walcott, "Mr. Carnegie's Gift to the Nation," *Independent* 53 (December 19, 1901): 2988. "Trust Deed by Andrew Carnegie," *The Carnegie Institution of Washington, D.C.: Founded by Andrew Carnegie* (Washington, D.C.: Press of the New Era Publishing Co., 1902), p. 11. On the national university issue, see the editorials, "Not a University," *New York Times*, January 10, 1902, p. 6, and "The Carnegie Institution," *New York Daily Tribune*, January 11, 1902, p. 6. See also "Finding the Exceptional Man," *Independent* 54 (November 27, 1902): 2847–49; "The Carnegie Institution," *Nation* 78 (January 14, 1904): 26.

6. Hugo Münsterberg, "St. Louis Congress of Arts and Sciences," *Atlantic Monthly* 91 (May 1903): 673. For praise of the congress, see Ernest Hamlin Abbott, "The Fair at St. Louis," *Outlook* 74 (July 4, 1903): 562, and Frederic C. Howe, "The World's Fair at St. Louis, 1904," *Cosmopolitan* 35 (July 1903): 286. Edwin E. Slosson, "A Clearing House of the Sciences," *Independent* 57 (October 6, 1904): 790–91. Millikan thought enough of Boltzmann's visit to record it forty-five years later in his *Autobiography* (New York: Prentice-Hall, 1950), pp. 84–85.

misunderstandings about science. The *Nation* considered the current public ideas of science to be medieval. For the common man, science was a "black art," Marconi was a "magician," Edison was a "wizard."[7] Science was identified by its theatricality and its inventions, its electrical lamp and its X-ray photography. The layman knew too little of the joys and disappointments of the ordinary researcher and appreciated less the meaning of his triumphs. Arthur G. Webster, speaking on Clark University's Founders' Day in 1907, complained that though the public's interest was aroused by great discoveries like radium, its continuing concern for science was negligible. Most large newspapers retained experts and critics for literature, music, and sports, but only three that Webster knew, the New York *Sun*, the *Evening Post*, and the *Boston Transcript*, did this for science. Though scientists reminded the public of their contribution to the general progress, the common man continued to believe that the scientists' work had no socially beneficial or "practical" consequences.[8]

The businessman, as well as the average citizen, was criticized for not appreciating science. W. G. Farlow made this condemnation: "If our business men are too stupid to take advantage of the help offered by science, although informed as to what is done by their foreign competitors, we shall not be called on to shed many tears over their ultimate failure in the competition for business."[9]

Farlow's assessment of American business was largely correct. American industry was slow to follow German competitors into basic and applied research. The laboratory of Höchst near Frankfurt-am-Main had been founded in 1863, that of the Badische Anilin und Soda-Fabrik of Ludwigshafen in 1865, but it was not

---

7. "Popular Appreciation of Scientists," p. 46. As late as 1911, the *Scientific American* was compelled to criticize the manner in which newspapers played upon the "miraculous" in science news. Editorial, *Scientific American* 105 (October 21, 1911): 362.

8. Arthur Gordon Webster, "America's Intellectual Product," *Popular Science Monthly* 72 (March 1908): 201–02. E. J. Townsend, "Science and the Public Service," *Science,* n.s. 32 (November 4, 1910): 613. This assumption about the common man's beliefs was also expressed in Lyman Beecher Stowe, "Patriots in the Public Service," *Outlook* 92 (July 24, 1909): 717.

9. Farlow, "The Popular Conception," p. 6.

until 1902 and 1903 that Du Pont established the Eastern Laboratory and the Experimental Station. American Telephone and Telegraph Company's subsidiary laboratories, the engineering department of Western Electric and the Bell Telephone Laboratories, had been founded in the nineteenth century, but not until 1907, when John J. Carty was appointed chief engineer, did the laboratories move into basic, electrical researches. General Electric had established its Standardizing Laboratory in 1895, but only in 1900 did Charles P. Steinmetz convince the company to allow him to establish an applied research laboratory. Kodak Research Laboratories was not organized until 1912. The Mellon Institute of Industrial Research was not founded until 1913. The greatest growth of industrial research laboratories came in the 1920s, when the number grew from three hundred in 1920 to one thousand in 1927.[10]

Arthur Webster attributed American success in industry to superior organization rather than to industrial science. The admonition was repeated that competition with foreign and especially German enterprises would require greater efforts in science.

10. References for the text are W. H. G. Armytage, *The Rise of the Technocrats: A Social History* (London: Routledge & Kegan Paul, 1965), p. 87; William S. Dutton, *Du Pont: One Hundred and Forty Years* (New York: Charles Scribner's Sons, 1942), pp. 180, 185; N. R. Danielian, *A.T.&T.: The Story of Industrial Conquest* (New York: Vanguard Press, 1939), pp. 92, 98, 102; John Anderson Miller, *Workshop of Engineers: The Story of the General Engineering Laboratory of the General Electric Company, 1895–1952* (Schenectady, N.Y.: General Electric Co., 1953), pp. 3, 17–19; Armytage, *Rise of the Technocrats*, p. 245; Edward R. Weidlein and William A. Hamor, *Glances at Industrial Research During Walks and Talks in Mellon Institute* (New York: Reinhold Publishing, 1936), pp. 17–27; Armytage, *Rise of the Technocrats*, p. 246.

There is no general history of American industrial research laboratories. Specific comments can be obtained in: W. H. G. Armytage, *Rise of the Technocrats;* Kendall Birr, *Pioneering in Industrial Research: The Story of the General Electric Research Laboratory* (Washington: Public Affairs Press, 1957); John Thomas Broderick, *Willis Rodney Whitney: Pioneer of Industrial Research* (Albany, N.Y.: Fort Orange Press, 1946); Courtney Robert Hall, *History of American Industrial Science* (New York: Library Publishers, 1954); Maurice Holland with Henry F. Pringle, *Industrial Explorers* (New York: Harper & Brothers, 1928); Kenneth E. Trombley, *The Life and Times of a Happy Liberal: A Biography of Morris Llewellyn Cooke* (New York: Harper & Brothers, 1954); Edward R. Weidlein and William A. Hamor, *Glances at Industrial Research;* and Weidlein and Hamor, *Science in Action: A Sketch of the Value of Scientific Research in American Industries* (New York: McGraw-Hill Book Co., 1931). See also Daniel Kevles, "The Study of Physics in America, 1865–1916" (Ph.D. diss., Princeton University, 1964), ch. 7. Kevles emphasizes the growth of industrial research before the war, whereas I obviously do not.

Superior organization was not necessarily the result of a manage-rial science, however, since business strongly opposed even this "science." Thus, Frederick W. Taylor's efforts to reorganize Bethlehem Steel Company on the principles of scientific manage-ment were resisted by management and finally disowned.[11]

These concerns for common ignorance and industrial indiffer-ence did not, however, move the scientists to popularize science. On the contrary, the professionalization of science culminating in these years further isolated the scientists and made popularization an unacceptable scientific activity. This was revealed, for exam-ple, in an episode concerned with one effort at popularization. In a *Century Magazine* article, George M. Stratton, professor of experimental psychology and director of the Psychological Labor-atory at Johns Hopkins University, argued that many railway disasters at night were caused by the use of colored lamps for signaling. He maintained that at night even people with normal color vision could not always distinguish colors and made recom-mendations to compensate for this deficiency in railway signal-ing. The whole argument was placed in the context of progressive reform with the *Century* editors calling for a commission of experts to investigate the accident problem. Published responses did not agree. Borrowing another writer's phrase, J. W. Baird, psychologist at the University of Illinois, consigned Stratton's article to " 'the class of antiquated and the non-scientific litera-ture.' " Baird contradicted the argument that normal color vision was deficient at night.[12]

11. Webster, "America's Intellectual Product," pp. 205–08. The theme of science's relation to industry was repeated also in E. J. Townsend, "Science and Public Service," p. 613; W. A. Hamor, "The Value of Industrial Research," *Scientific Monthly* 1 (October 1915): 86–92; and T. Brailsford Robertson, "The Cash Value of Scientific Research," *Scientific Monthly* 1 (November 1915): 140–47. For an account of the Taylor episode, see Frank Barkley Copley, *Frederick W. Taylor: Father of Scientific Management*, 2 vols. (New York: Harper & Brothers, 1923), II, ch. 11; also see Samuel Haber, *Efficiency and Uplift: Scientific Management in the Progressive Era, 1890–1920* (Chicago: University of Chicago Press, 1964), p. 35.

12. George M. Stratton, "Railway Disasters at Night," *Century* 74 (May 1907): 118–23. "Safety on the Railroads," *Century* 74 (June 1907): 321. J. W. Baird, " 'Popular' Sci-ence," *Science*, n.s. 26 (July 19, 1907): 75–76. Stratton replied to Baird's criticisms in "Railway Accidents and Color Sense," *Popular Science Monthly* 72 (March 1908): 244–52.

The significance of this exchange was not that scientists disagreed about facts. Such disagreement has not been infrequent in the history of science. Rather, the controversy illustrated the professional unacceptability of the popularization as a contribution to the progress of science. Stratton's essay was in the class of "antiquated" and "non-scientific" literature. Baird did not assert that it was unscientific; he asserted only that it was not the sort of article that scientists any longer wrote. Stratton's errors of fact would not alone have drawn Baird's criticism that the article was not professional. But errors of fact contained in a popular publication, rather than in a professional paper, did make them nonscientific. The popular article was no longer part of professional science, even if it was authored by a professional scientist.

Professional disdain for popularization was a shift away from the traditions of the nineteenth century. Popular science was then, by and large, an acceptable contribution to the advance of science. The popular lecturing of Faraday, Helmholtz, Huxley, and Tyndall indicates this. Popularization also played an important role in the struggle for acceptance of the Darwinian theories.

Although there is no entirely convincing theory for the origin and support of nineteenth-century popularization, one historian has suggested that the public acceptance of professional science led to the professional exclusion of popularization. Specialization and professionalization of science before the Civil War withdrew increasing amounts of general knowledge from the public domain; thus natural philosophy and natural history, pursued by amateurs and naturalists, were specialized into physics, chemistry, geology, meteorology, zoology, botany, and other sciences. The esoteric quality and the professional research standards of these sciences prevented many older scientists, as well as naturalists and other educated persons, from following daily scientific progress. Popularizing was the natural response for the specialists to make in a democratic society to justify specialization and to end public confusion over what they were doing.[13] In the professional

---

13. George H. Daniels, *American Science in the Age of Jackson* (New York: Columbia University Press, 1968): pp. 40–41.

struggle over Darwin's theories, for example, knowledge being withdrawn from public domain included the facts, reasoning, and theories of the creationist view of geology and biology that gave evidence for the existence of a deity and his plan for the world. Darwinian biology invalidated the analogical reasoning, reinterpreted the facts, and denied the conclusions of scientists, like Louis Agassiz, whose work had been supported by the public. Popularization was necessary to explain the position of the new science on the old beliefs. It was to be expected, therefore, that when professional science had become accepted—or was no longer attacked—by the public, this motivation would disappear.

Scientists had also popularized their work to obtain social and financial support from the educated middle class. In the late nineteenth and early twentieth centuries, other sources of support were found in industries, universities, and philanthropies, which generally could understand science in its own terms or knew its value for their own purposes. Consequently, popularization declined as a professional activity.[14]

By 1900, popular science had become a literary and genteel indulgence. William J. Humphreys, meteorological physicist for the weather bureau, described the need for popular science in that popular science offered "that knowledge and understanding that broadens our sympathies; that increases our interest in the world around us; that makes us more contented and more useful human beings." There was "no nobler work" than popularizing science.[15] This was an admirable sentiment, but as a description of the value of popular science, it contained nothing to distinguish it from the reputed value of travel books and genteel fiction. That popular science was placed in this class of literature goes far to explaining why prewar criticism of popularizations was more concerned with artistry and craftsmanship than with scientific content.[16]

No one doubted that popularization was in precipitant decline.

14. Daniel J. Kevles, "The Study of Physics," pp. 257–68.

15. W. J. Humphreys, "Right and Wrong in Popular Science Books," *Science*, n.s. 29 (February 19, 1909): 297.

16. For example, the editorial, *Scientific American* 105 (October 21, 1911): 362.

But as a community, the scientists had only mild misgivings and did not move to halt it. In 1900, James McKeen Cattell, the editor of *Popular Science Monthly*, wrote to the leading American astronomer, Simon Newcomb, for aid in popularizing science. Cattell knew the great scientist was not enthusiastic about the need for popularization and wrote defensively to forestall criticism: "Perhaps I overestimate the importance of maintaining direct relations between the scientific worker and those who are not directly engaged in scientific research but I think that you will at all events agree with me that this is not unimportant." It was not, however, considered important. Three years later, Cattell had to write again to Newcomb, this time to complain that the Smithsonian Institution was doing "absolutely nothing with the Smithson Fund for the promotion and diffusion of knowledge." The *Nation* commented that specialization and mathematics seemed to have driven popular interest from science. Five years later, the disreputability of newspaper scientific news impelled the *Nation* to make an unanswered plea for a new effort at popularization. In 1913, Edwin Slosson, speaking before the School of Journalism of Columbia University, complained that newspapers did not give scientific events the coverage they deserved. Indeed, they often treated science with contempt. This same year, Cattell would write discouragingly to a friend, "Neither am I hopeful about the likelihood of popularizing science or extending its influence among the public."[17]

By 1915, the nineteenth-century tradition of popular science had ended. Scientists were vaguely esteemed, though they complained of being misunderstood and not appreciated. But the profession had excluded popularization as a valid scientific activ-

17. James McKeen Cattell to Simon Newcomb, May 9, 1900, box 19, Simon Newcomb Papers, Manuscript Division, Library of Congress. Cattell to Newcomb, January 31, 1903, box 19, Newcomb Papers. Nathan Reingold has suggested, in a private communication with the author, that the quoted passage in this letter referred to sponsorship of research as well as to popularization. "Exit the Amateur Scientist," *Nation* 83 (August 23, 1906): 159–60. "Scientists and the Masses," *Nation* 92 (May 4, 1911): 441. Edwin E. Slosson, "Science and Journalism: The Opportunity and the Need for Writers of Popular Science," *Independent* 74 (April 24, 1913): 914. Cattell to William H. Welch, April 26, 1913, box 10, George Ellery Hale Papers, The Carnegie Institution of Washington and the California Institute of Technology, Pasadena, California.

ity. Only a few scientists, like Cattell, condemned the failure of a democratic society to promote popular understanding of science. Even the name *popular science* was in disrepute. This was a double loss because popular science had the task of conveying the value of science to the public. Individual and library subscriptions were insufficient in 1915 to support the great journal *Popular Science Monthly*. The magazine's publishers had been losing ten thousand dollars a year and decided to cease publication.[18] Henceforth, daily scientific research would be discussed in the new journal, *Scientific Monthly*, with a narrow and mainly scientific readership. The magazine bearing the name *Popular Science Monthly* would eventually be devoted to popular technology, gadgets, and hobbies.

The Rise of the Experts

In evident contrast to declining interest in scientific popularization and the cultural isolation of professional, pure science, science was given significant social value in progressive liberalism by Herbert Croly and Walter Lippmann. For both men, science was one key to solving the chief social problem of their time, the unequal distribution of wealth that disrupted American unity.

In *The Promise of American Life*, Croly's now famous solution to the social problem was the use of Hamiltonian methods to achieve the Jeffersonian goal, or national organization of society under the leadership of an elite to achieve social and economic democracy. Popular adherence to traditional individualism (and laissez faire), however, prevented acceptance of the Hamiltonian methods. One of the central efforts of social reformers was therefore to redefine the concept of individualism. Real individuality, Croly wrote, was not achieved by merely doing well what everyone must do to live, such as making money; individuality consisted in doing with ability, energy, disinterestedness, and excellence what no one else could do. Individuality brought personal

18. [James McKeen Cattell], "Scientific Journals and the Public," *Popular Science Monthly* 87 (September 1915): 309.

(that is, unique) distinction. It fully developed the special characteristics, traits, and talents of one's own personality.

Individuality would not contribute to national cultural unity,
however, unless it demanded excellence and disinterestedness.
Judgment of performance required accepted standards. Here
Croly turned to science. Science, he thought, provided the best
examples of these high standards: "The perfect type of authoritative technical methods are those which prevail among scientific
men in respect to scientific work. No scientist as such has anything to gain by use of inferior methods or by the production of
inferior work."[19] Croly considered this an ideal model for the
democratic society in which every individual would be expected
to make a contribution. The large variety of arts, crafts, and sciences, with their many schools and disciplines, could allow everyone an arbitrary decision toward an endeavor suited to his personality.

Science was important not simply because it provided a model
for the reform of society. Scientists themselves should be included
in the leadership elite. It was incumbent upon scientists as intellectuals to make America aware that the social problem would
not be alleviated by traditional means. They also had to convince
the people of the efficacy and legitimacy of the Hamiltonian
methods. By their own intellectual performance, they had to
demonstrate that authoritative standards could bring progress
and unity to a culture. It was then necessary for these standards
to be popularized and the common man committed to them.[20]

This view of the decisive place of science and the role of its
practitioner had its strongest exercise in Lippmann's *Drift and
Mastery*. Science would save democracy by mastering all economic, political, and social problems. "Rightly understood science is the culture under which people can live forward in the
midst of complexity, and treat life not as something given but as
something to be shaped." Democracy was the political version of

19. Herbert Croly, *The Promise of American Life* (1909; reprint ed., New York: E. P.
Dutton, 1963), p. 434.
20. Ibid., pp. 434, 444.

the scientific method. Science would halt social disintegration and provide cultural unity because it was "not alone a binding passion, but a common discipline." Science would promote the necessary cooperation to solve large problems. To assuage the fears that science was inhuman and removed from the average man, Lippmann argued that science sprang from the basic human need not only to understand but to control the physical world.[21]

The apparent paradox raised by the central role of science in progressive social thought can be resolved only when the major developments bringing science to this position are pointed out. The first, certainly, was the absorption of Darwinism into liberalism. The Spencerian development hypothesis of the 1850s and the Darwinian theory of evolution in the next decade had provided a rationalization for the spectacular growth of industries and personal fortunes in the Gilded Age and provided, as well, a conservative defense against the efforts of reformers to meliorate the resulting social conditions. Combining the Protestant Ethic, classical economics, and the theory of natural selection, social Darwinism interpreted society in terms of the contractual relations of freely competing individuals. This form of society was thought to be the only one naturally capable of progress. Social Darwinism was the "steel chain of ideas," as Eric Goldman has said, which protected the new industrial order.

Though Darwinian science was a key link in this chain of ideas, it was a reinterpretation of Darwinism that broke the chain and provided the philosophical basis for social reform. Liberals reasoned that an accurate understanding of Darwinian theory did not justify the status quo; to the contrary, it indicated that continual change was the primary characteristic of man and nature. Planned manipulation of men and the physical environment could direct the course of evolution. This intellectual liberation was carried into most aspects of modern thought, from the social gospel and the new economics to psychology and philosophy.

21. Walter Lippmann, *Drift and Mastery* (1914; reprint ed., Englewood, N.J.: Prentice-Hall, 1961), pp. 151, 154, 165.

These ramifications were all united in their genetic history and their assumption of man's creative freedom.[22]

The place of science in progressive philosophy was indebted not only to Darwinism but also to the concept of the expert. This concept was to leave an ambiguous legacy for the scientists in the 1920s. The "expert" gave the clearest expression of the role in society of the man with specialized knowledge and provided the obvious analogy for the scientists in making their claim to social value. Yet the concept of the expert excluded reference to the pure scientist. This can be illuminated by examining the three influences on the progressive concept of the expert: scientific government, scientific management, and professional engineering.

The most important statement of the progressive theory of the expert was Charles McCarthy's *The Wisconsin Idea* (1912). One problem in achieving good government was inefficiency and fraud in the administration of laws. Progressives considered the formation of commissions a method of extending legislative power to the daily regulation of business. But regulatory commissions would be ineffective if their staffs had insufficient administrative experience and lacked intimate knowledge of the businesses being regulated. Only if commissioners and staffs were trained in administrative science, economics, and sociology could the commission be effective. Science was to be a cure and a substitute for corrupt government. More particularly, scientific administration, with the ideals of duty, permanence, and scientific method, would remove the power of interest groups.[23]

22. The standard literature on the development of modern liberalism includes Charles Forcey, *The Crossroads of Liberalism: Croly, Weyl, Lippmann, and the Progressive Era, 1900–1925* (New York: Oxford University Press, 1967); Eric Goldman, *Rendezvous with Destiny: A History of Modern Reform* (New York: Random House, Vintage Books, 1956); Richard Hofstadter, *Social Darwinism in American Thought* (Boston: Beacon Press, 1955); David Noble, *The Paradox of Progressive Thought* (Minneapolis: University of Minnesota Press, 1958); and Morton White, *Social Thought in America: The Revolt Against Formalism* (Boston: Beacon Press, 1957). Dates of publication are for paperback editions, if these have been issued.

23. For the main lines of this discussion, but not the conclusions, I am indebted to Rush Welter, *Popular Education and Democratic Thought in America* (New York: Columbia University Press, 1962), pp. 258–60. Charles McCarthy, *The Wisconsin Idea* (New York: Macmillan, 1912), pp. 178, 186. This was the import of Lippmann's statement that the scientific method worked against one's interest, see Lippmann, *Drift and Mastery*, p. 165.

Science was also to make the legislative process less corruptible and more efficient. Legislators often did not know enough to write a bill, made misguided attempts to adapt another state's law, or were unaware of previous laws relevant to a proposed law. McCarthy's solution to these difficulties was the legislative reference service, a department consisting of trained librarians and experts in the areas of the legislature's standing committees.[24]

For McCarthy, the expert was scientifically trained in political and social sciences, administration, and economics. He had specialized experience and had applied his formal theories and tested them. He had common sense. The expert was eminently a practical man.

The goal of the expert was efficiency. In the progressive era, efficiency meant not simply effectiveness and lower costs, but disinterestedness and professionalism. It carried a sense of moral approval. Efficiency had become an ideal of the progressives by the time McCarthy published his work. Its popular impact was made in 1910–1911 in the Eastern Rate Case, initiated by the railroads' demand for an increase in rates from the Interstate Commerce Commission. Representing eastern businessmen and small shippers opposed to the increase, Louis D. Brandeis produced testimony for the argument that if the railroads became more efficient they would not need to raise rates (to maintain their level of profit). A national efficiency craze developed, sustained in 1911 by a congressional investigation which brought before it Frederick W. Taylor and his scientific management people.[25]

The concept of the expert in government and the concept of efficiency were reinforced by the emergence of professionalism and a sense of public responsibility in the engineering community. Just as the middle class came to believe that the technical expert could save democracy, the professional engineer began to

24. McCarthy, *The Wisconsin Idea*, p. 186.
25. For a discussion of the meanings of efficiency, see Samuel Haber, *Efficiency and Uplift*, Introduction and ch. 4.

publicize his utility and responsibility to the social welfare.[26]

The decisive turning outward from the tradition of craft and professional concerns came in 1908 with Morris L. Cooke's address before a meeting of the American Society of Mechanical Engineers, "The Engineer and the People: A Plan for a Larger Measure of Cooperation Between the Society and the General Public." Cooke urged an educational campaign to acquaint the public with the social utility of engineering so that the engineer could serve the public interest directly. He recommended publicity committees, public lectures at the Engineering Societies building (completed in 1907 in New York City), and news dispatches to newspapers. He suggested incidentally that the internal progress of the profession itself depended on fulfilling the profession's public responsibility.[27] The pure scientists certainly did not hold a similar view at this time.

The discussion following the address included the comment that the public should appreciate engineering because it had made the modern world. The mayor of Pittsburgh, George W. Guthrie, thought engineers could remedy public evils. Indeed, engineers were in the forefront of the city-manager movement because their technical knowledge of utilities, for example, was necessary to remove fraud and to institute honest, public administration. Frederick W. Taylor, a former president of the society, Ambrose Swasey, the magnate of the machine-tool industry, and Arthur Hadley, president of Yale University, endorsed Cooke's recommendations.[28] Their presence at this meeting indicated the

26. For a discussion of the engineer's sense of public responsibility at this time, see Monte A. Calvert, *The Mechanical Engineer in America, 1830–1910* (Baltimore: Johns Hopkins Press, 1967), ch. 14, especially pp. 271–72; and Edwin Thomas Layton, "The American Engineering Profession and the Idea of Social Responsibility" (Ph.D. diss., University of California, Los Angeles, 1956), ch. 5. Layton attributes the engineer's social sense only to the general influence of the progressive movement.

27. Morris Llewellyn Cooke, "The Engineer and the People: A Plan for a Larger Measure of Cooperation Between the Society and the General Public," *Transactions, American Society of Mechanical Engineers* 30 (1908): 619–28, 637. Trombley's biography of Cooke, *The Life and Times of a Happy Liberal*, does not deal with this aspect of Cooke's early career.

28. Abstracts of the comments in this discussion were printed following Cooke's article, "The Engineer and the People," pp. 628–37.

deep implication of the engineer in the efficiency craze and the progressive movement.

As a consequence of the association in the public mind of the engineer's social responsibility, the expert in government, and scientific management's ideal of efficiency, the progressive concept of the expert was given connotations of practicality and technical knowledge. This prevented the pure scientist from being considered as an expert. It was not alone the possession of knowledge and the scientific method which distinguished the expert, but the problems to which these were applied. In the public mind, the expert had preempted all claims to "scientific usefulness," thus accentuating the apparent nonutility of the pure scientist. This preemption was probably the main reason the public believed that the scientist's search for truth about the natural world did not contribute to the improvement of the average man's daily life.

The preemption had the effect of excluding the pure scientist from the progressive concept of science itself. When the progressive spoke of the scientist, his primary reference was to the impartial expert who had the practical knowledge and ability to manipulate the environment, to make democracy work. In the new urban age, the function of science was to generate methods or rules which would guide successful and efficient adjustment of conflicting claims to privilege. Thus scientific management and the rule of efficiency originated in the efforts of scientifically trained hydrographers and foresters, like Gifford Pinchot, to develop water resources and forests for maximum, long-term use and to protect them from destruction by special, short-term interests. Industrial scientific management was to increase productivity and unify the interests of laborers and employers. Rationalizing methods provided executive administrators with means to respond quickly and objectively, that is, efficiently, to erupting social problems. The scientist was not a man in a laboratory. As one historian has said, "The sense of science not simply as a means of organizing knowledge or wisely viewing the universe but as a method for getting things done, exploring the practical problems of human experience, was central to progressive thought."[29] This concept of science would be an obstacle to the efforts of scientists in the 1920s to relate their values to progressivism, to promote cultural unity, and to organize peacetime scientific work.

29. Barry Dean Karl, *Executive Reorganization and Reform in the New Deal: The Genesis of Administrative Management, 1900–1939* (Cambridge, Mass.: Harvard University Press, 1963), p. 67. The progressive concept of science is explored in Haber, *Efficiency and Uplift;* Samuel Hays, *Conservation and the Gospel of Efficiency: The Progressive Conservation Movement, 1890–1920* (Cambridge, Mass.: Harvard University Press, 1959); Karl, *Executive Reorganization and Reform;* and Robert H. Wiebe, *The Search for Order, 1877–1920* (New York: Hill and Wang, 1967).

(2) THE AWAKENING OF THE SCIENTISTS, 1916-1920

During the First World War, nongovernmental scientists in universities and research institutions were recruited to work on defense problems. Their scientific activity in the war was distinctly different from their earlier work. Before the war, they had done research on problems whose solutions were of interest mainly to men of their own specialities. These researches had been individual enterprises in which they had worked without supervision. Their professional activities had been conducted on the local or regional levels except for the annual or semi-annual national conference in their fields. In contrast, during the war many men left their homes for research centers like the New London Experimental Station or Washington, D.C. Their research was a team effort, supervised and coordinated with that of other teams by a central agency. And these scientists had the deep satisfaction of knowing that they contributed directly to America's survival.

The war work had an important impact on scientists. Their experiences lessened antagonism toward centrally administered research and toward mission research at the national level. The self-sufficiency of specialized professional life was questioned. And, more importantly for this study, the scientists became concerned that their professional values should correspond with the broader economic and political values of the American people.

The Isolation of Professional Scientists

The scientists' attitudes before the war were revealed in their responses to George Ellery Hale's attempt to reform the National

Academy of Sciences. Chartered in 1863 as a private organization for consultation by the government on scientific and military questions which arose in the Civil War, the National Academy of Sciences was, as well, an attempt by the luminaries of science, Alexander Dallas Bache, superintendent of the Coast Survey, Charles H. Davis, chief of the Bureau of Navigation, Joseph Henry, and Louis Agassiz, to centralize control over American science.[1] The academy accomplished neither of these objectives. The growth of governmental scientific bureaus usurped its advisory function. The sponsorship of science in state educational institutions, private universities, and philanthropic institutions, like the Carnegie Institution and museums, prevented the academy from centralizing science. After 1900 the academy was moribund, enrolled mostly older scientists who were no longer productive, and had little scientific function within the profession. The obscurity of the academy was so great that by the time of the preparedness controversy of 1915 its members had to remind President Wilson of its chartered role.[2]

George Ellery Hale was the director of the Mount Wilson Observatory. He had been one of the founders of astrophysics, a science that infused experimental physics into an American tradition of observational astronomy. He established the *Astrophysical Journal* (1894), the Yerkes Observatory, and later the Mount Wilson Observatory. He helped found the California Institute of

1. Ralph S. Bates, *Scientific Societies in the United States* (Cambridge, Mass.: M.I.T. Press, 1965), pp. 77-84; Thomas Coulson, *Joseph Henry: His Life and Work* (Princeton: Princeton University Press, 1950), pp. 277-81, for Henry's ambivalence toward establishment of the academy; A. Hunter Dupree, *Science in the Federal Government: A History of Policies and Activities to 1940* (New York: Harper & Row, Torchbooks, 1964), pp. 115-19, for a discussion of Bache's ideas on centralization, and pp. 135-48; Edward Lurie, *Louis Agassiz: A Life in Science*, abridged ed. (Chicago: University of Chicago Press, 1966), pp. 331-36; Frederick W. True, ed., *A History of the First Half-Century of the National Academy of Sciences, 1863-1913* (Washington, D.C.: National Academy of Sciences, 1913), pp. 1-23.

2. Edwin Grant Conklin to President Wilson, October 21, 1915, box 11, George Ellery Hale Papers, The Carnegie Institution of Washington and the California Institute of Technology, Pasadena, California (cited hereafter as Hale Papers). A personal meeting between Hale and other academy members and President Wilson was necessary in April 1916 to impress the president with what the academy could do in wartime. See Helen Wright, *Explorer of the Universe: A Biography of George Ellery Hale* (New York: E. P. Dutton, 1966), p. 287.

Technology and the Henry E. Huntington Library and Art Gallery. All his innovating energy was brought to the reform of the National Academy of Sciences.[3] Though his mature vision of a revitalized academy would not come for another decade, his election to the academy in 1902 led him to think in the terms of general knowledge which guided his later efforts. He wrote to Harry Manley Goodwin, a friend since his student days at the Massachusetts Institute of Technology, that he was "becoming more and more interested in investigation in general, regardless of particular field of knowledge."[4]

Hale's analysis of the role of the academy and his prescription for reform were elaborated in his series of articles (from 1913 to 1915), "National Academies and the Progress of Research." Surveying the historical role of academies, he not unexpectedly judged that they had been a major influence in stimulating individual scientists, promoting research by provision of grants and laboratories, popularizing science, and providing a sense of community among scientists. Their greatest achievement had been to dignify and honor the pursuit of research and to promote appreciation for research among nonscientists. Like the founders of the National Academy of Sciences, he concluded that if an academy was to flourish, it had to have cooperation from the government. Nevertheless, Hale looked less to the state-sponsored L'Académie des Sciences as a model for the National Academy than to the Royal Society of London. The Royal Society received little state patronage and maintained no laboratories of its own. Despite the traditional individualism of British scientists, the Royal Society was able to promote the progress of British science by extending membership and honors, bringing scientists together at meetings, and publishing the *Philosophical Transactions*. With conditions similar to those in England, the National

3. Helen Wright's biography of Hale, *Explorer of the Universe*, admirably recalls his private life, though it does not emphasize the significance of his political and ideological activities. See chs. 14 and 15.

4. Hale to Harry Manley Goodwin, May 19, 1902, box 2, HM 28456, George Ellery Hale Collection, Henry E. Huntington Library and Art Gallery (hereafter cited as Hale Collection). This item is quoted by permission of the Huntington Library, San Marino, California.

Academy should also be able to achieve similar results.[5]

Hale believed that only the National Academy could properly represent American science in international assemblies and unions.[6] He abhorred scientific provincialism and wanted American science to assume leadership in cooperative, international science.

"The Future of the National Academy of Sciences" contained Hale's recommendations for the reform of the academy. Not published until December 1914, the article was, however, delivered as an address at the Baltimore meeting of the academy in November 1913. In general, he recommended that the academy should have a home building and its own laboratories, support original research, and establish a working relationship with the national government. To the members, it should contribute financial and laboratory assistance. To industry, it should demonstrate the dependence of applied science on pure science. To the public, the home building should be "visible evidence of the Academy's existence." *Proceedings* should be published as the official journal, carrying abstracts and reviews of original research to all scientific institutions abroad, thereby raising American prestige. He realized that most scientists avoided popular publicity, but he thought the distance between the scientist and the layman had to be bridged. He recommended that abstracts of scientific papers be distributed to responsible, nonscientific magazines for the educated public. Public lectures should be given. The membership of the National Academy should be increased to include young scientists whose current research would enhance the standing of the academy and to make it more representative of American science.[7]

5. George Ellery Hale, "National Academies and the Progress of Research," *Science*, n.s. 38–40. The individual articles are: "I. The Work of the European Academies," n.s. 38 (November 14, 1913): 681–98; "II. The First Half Century of the National Academy of Sciences," n.s. 39 (February 6, 1914): 189–200; "III. The Future of the National Academy of Sciences," n.s. 40 (December 25, 1914): 907–19.
"National Academies and the Progress of Research: II," *Science*, n.s. 41 (January 1, 1915): 12–23.
6. Hale, "National Academies and the Progress of Research, II," p. 199.
7. Hale, "National Academies and the Progress of Research, III," pp. 910–11, 914–17; "National Academies and the Progress of Research: II," pp. 12–19. For a discussion of

There were two objectives in these proposals: Hale wanted not only to revive the National Academy and to advance science professionally, but also to create a national science. National science for Hale had several related connotations. It meant, first, the centralized organization of scientific activity along interdisciplinary lines on the national level (although he did not want the academy to have direct authority over individual research). National science also meant the elevation of science to a prominent influence in American culture. For Hale, such influence was symbolized by the official prestige of the European national academies.

Both the organizational objective and the cultural objective of the National Academy reform proposals were clearly revealed in Hale's letters from 1912 to 1914 to Sen. Elihu Root and Andrew Carnegie. Hale had been acquainted with Root since the early 1900s when both became attached to the Carnegie Institution, Root as a trustee and Hale as director of the Mount Wilson Observatory. Root was to emerge as one of Hale's "most valued and beloved advisers."[8] Hale naturally turned to Root in 1912–1913 with his plans for the National Academy of Sciences's home building. Root, in consultation with Henry Pritchett, also a Carnegie Institution trustee, urged Hale to make his scheme ambitious. It could more easily find funds by being trimmed than expanded.[9] In two long letters to Root in March 1913, Hale discussed his proposed reforms and explained how a revived national academy could take a role in nonscientific culture.

In the first letter, he admitted the disadvantages for a national academy in a country lacking a natural intellectual center. But

---

the *Proceedings*, see A. A. Noyes et al., "Report of the Special Committee on the Publication of a Journal by the Academy," *Report of the National Academy of Sciences for the Year 1913* (Washington, D.C.: Government Printing Office, 1914), pp. 25–27; Hale, "The Proceedings of the National Academy of Sciences as a Medium of Publication," *Science*, n.s. 41 (June 4, 1915): 815–17; also Edwin Bidwell Wilson, *History of the Proceedings of the National Academy of Sciences, 1914–1963* (Washington, D.C.: National Academy of Sciences, 1966).

8. Wright, *Explorer of the Universe*, p. 308. For Root's autobiographical testimony to personal interest in science, see Philip C. Jessup, *Elihu Root*, vol. 1, *1845–1909* (New York: Dodd, Mead, 1938), pp. 278–79.

9. Root to Hale, February 17, 1913, box 35, Hale Papers.

this difficulty could be obviated. An academy journal would allow members to announce the results of research through a national body which would place recognition of the academy on the same level as other national societies. A journal would also allow the academy to address itself to the larger problems of science. Hale wrote that in previous years the academy had "exhibited a narrow tendency, and a real disinclination to take a broad attitude toward the scientific interests of the country at large." That spirit was hopefully disappearing.[10]

In his second letter, Hale was primarily concerned with the cultural objective. National science should be devoted to three matters: cooperation in international science, the public appreciation of science, and a role in humanistic studies and arts. Professional scientists would not consider any of these as necessary to the progress of science in America. They might only concede that the first item could be important.

An academy building, Hale told Root, would allow the academy to host the International Association of Academies, thus making the National Academy of Sciences the " 'leading Academy' " of the association. A building would allow greater participation in cooperative international research. It would raise the prestige of American science. Appropriate exhibitions in the building could impress on American industry the benefits of applied science derived from the cultivation of pure science. He stressed that if American industry was to compete successfully with German industry, it must undertake pure research.

He was disturbed that science was not considered a cultural subject in the college curriculum. Students of the arts resented having to learn about pure science which had no relevance to American culture. Students of the sciences resented having to learn about pure science which had no direct utility in their professions. And technical schools taught the sciences as a profusion of details without any unifying ideas. Even professional scientists failed to recognize the place of their specialty in the broad field of scientific activity. The humanists suffered from this same intel-

10. Hale to Root, March 3, 1913, ibid.

lectual impoverishment. Hale had received personal testimony from learned nonscientists who regretted their lack of a general view of science. He thought these educational deficiencies could be filled by the study of evolution, including cosmic, biological, and cultural evolution. This study would present a whole picture of nature. It would inspire interest in science. But its cultural value was most important: "It would stimulate the imagination no less profoundly than the best works of art or literature." It would end the estrangement of the humanities from the sciences by demonstrating evolution's role in all human activity. "And it would aid in breaking down the provincial narrowness of the man whose mind is wholly preoccupied with the details of his own life." The National Academy could assist this educational effort by public lectures, exhibitions, and a model museum of evolution. This could not fail to have an effect on the public.[11]

Hale rehearsed and added to this argument for an academy building in a letter to Andrew Carnegie. The National Academy, "because of its representative and truly national character," could promote all scientific fields. It could offer guidance for local scientific societies. Science in America should serve national interests as well as professional interests.[12]

At the November 1913 meeting of the National Academy, an unpublished draft of Hale's paper on the future of the academy was read in his absence by Edwin Grant Conklin. Following the meeting, Arthur L. Day, home secretary, had copies of the paper distributed to all the members with the request for their comments on the proposed reforms.[13] The fact that Hale felt it necessary to make the recommendations indicated the torpor of the academy. The responses of the members indicated more precisely the temper of the membership and, by inference, that of the whole scientific community.

Arthur Day prepared a numerical summary of the members' replies. Of the 132 domestic members of the academy, only 75

11. Hale to Root, March 10, 1913, ibid.
12. Hale to Andrew Carnegie, May 3, 1914, box 10, Hale Papers. Hale's efforts to finance the academy building and the composition of this letter to Carnegie are given fuller elaboration in Wright, *Explorer of the Universe*, pp. 308–12.
13. Hale, "National Academies and the Progress of Research, III," p. 907.

replied (approximately 57 percent). The table below indicates the division of opinion.

Recommendation For	In Favor	Opposed	No Opinion
Academy building	60	10	5
Strong relationship with federal government	3	1	71
Academy laboratories	6	14	55
Public lectures	26	10	39
"Proceedings"	35	17	23
Increase in membership	52	8	15

Of all the replies, only 15 members expressed general agreement with Hale's suggestions.[14] In short, "radical" reform was not endorsed.

Hale had been forewarned by Edwin Conklin that his reforms would not be popular. They would "undoubtedly stir up the Academy," Conklin wrote. "I only fear that your suggestions will prove to be too numerous and too radical for some of our conservative members." He tried unsuccessfully to persuade Hale to present his reforms piecemeal. But Conklin's reservations about the presentation of the ideas did not reflect his opinion of them. He also thought the academy needed a "new birth." It stood apart from the members' professional lives. It did not fulfill its statutory obligation to the government. It represented most clearly only the gulf between the public and science. Like Hale, Conklin thought that "the Academy touches entirely too few people, and in some way must be brought into more intimate relations with the life of the country at large."[15]

Some members commented rather curtly. Henry Fairfield Osborn, president of the American Museum of Natural History,

14. Arthur L. Day, "Summary of opinions of members re suggestions in Mr. Hale's paper: 'The Future of the National Academy of Sciences,' " November 27, 1914, box 53, Hale Papers.
15. Conklin to Hale, March 28, 1913, and November 14, 1913, box 11, Hale Papers.

commented, "I regret to say that I am not in sympathy with the majority of his suggestions." Some replies were more elucidative. William Dall of the Smithsonian Institution agreed that the academy should have a home building, but thought the necessary endowment could not be raised. Even more, he did not think the building could be useful to all the members. Only the Washington members would have continual access to it and it would turn into another local club. He doubted that the building would give the academy a professional function because the working scientific societies were always specialized and local. He doubted the usefulness of laboratories in the capital. Finally, Dall opposed cooperation with the national government out of a fear that bureaucracy would inhibit disinterested and free research.[16]

T. W. Richards, physicist and director of the Walcott Gibbs Memorial Laboratory at Harvard University, wrote that the geographical size of America and the impecuniosity of the scientist prevented the academy from playing an important role in the life of the researcher. C. L. Jackson, a chemist at Harvard, agreed with his colleague that America's size mitigated against an active academy. He confirmed Dall's opinion that the professionalization of science in America made general societies—except for the distinction of membership—superfluous.[17]

One of the more thoughtful letters was written by H. Fielding Reid, a geologist at Johns Hopkins. He reiterated the common objections that the country was too large and the scientific community too diffuse. He did not think that public lectures and exhibitions could serve any useful purpose. But Reid's misgivings about Hale's recommendations were deeper than these criticisms indicated. Hale's address implicitly carried a bias for utilitarianism (defined as the relevance of science to nonprofessional interests) and nationalism. Reid did not think that the academy should promote useful knowledge. It should encourage only purely scientific research. Its membership should be clearly distinguished from the membership of professions "mainly followed for pecuni-

16. Osborn to Arthur L. Day, December 18, 1913; Dall to Day, December 11, 1913, box 53, Hale Papers.
17. Richards to Day, December 11, 1913; Jackson to Day, December 14, 1913, ibid.

ary gain." The academy should stand for the scientific ideal alone. He resented the idea that the academy should foster special sciences in its laboratories. He was opposed to the academy's controlling or directing any research: the freedom of the researcher was absolute.[18]

Though the majority of replies were either hostile or indifferent, Hale remained optimistic about the future of his proposed reforms. In April 1914 at a business meeting of the academy, he attempted to counter the criticisms. This talk gave further indication of his goal of national science.[19] Hale did not agree, obviously, that the main purpose of the academy should be to confer distinction by membership. He even wanted the membership to include representatives of the humanities, though he had to calm fears that this would dilute the prestige of the academy as a scientific body.

The main criticism which Hale had to meet concerned the centralization of science. This issue was embodied in his proposal for academy laboratories. These would tend to centralize scientific research; they might attract foundation money which otherwise would be dispersed to universities; and if they were successful, they would exert pressure for centralization of facilities in other regions. Hale said that his scheme for laboratories had not been thought through, but as he elaborated the scheme it was perfectly plain that he had given it much consideration. He wanted two laboratories, one for physical science and one for biology, each with a secretary, research directors, and staff. By scientific discoveries and popular lectures the laboratories would add luster to the academy. He mentioned that one of the upper floors of the academy building might provide suites for the president of the academy or another prestigious scientist (Hale was not precise). His vision grew grander as he spoke. It seemed amenable to indefinite elaboration.

Hale did not convince the academy to accept his grand design. The adoption in 1914 of his recommendation to publish a journal

18. Reid to Day, December 18, 1913, ibid.
19. The following discussion is based on the "Proceedings of the Business Meeting of the Academy [apparently a stenographic typescript]," April 21, 1914, ibid.

left his other proposed reforms unadopted. In May, letters from Henry Pritchett and Andrew Carnegie denied his requests for money.[20] An early episode in reform ended.

The scientific community remained decentralized, its activities apportioned between the federal scientific bureaus, state scientific commissions, educational and philanthropic research institutions, a handful of industrial laboratories, and other institutions, like museums. The working scientific societies were small, local, and specialized. The leading scientists were indifferent to reform of this situation or were confirmed in the belief in its advantages. The cultural dichotomy between science and the rest of society, which had existed since the evolution controversy, was not considered harmful enough by the majority of scientists to motivate support for Hale's reforms. Scientists were confident of their professional success. Indeed, the new physics had since 1895 created new research opportunities. The scientists felt no need to move beyond their profession.

Professional isolation and cultural dichotomy prevented the creation of a national science. National science would have to do more than promote the professional reputations of individual scientists. It would have to enhance the prestige of American culture. The majority of scientists did not think the divergence of science from the rest of culture endangered professional progress; on the contrary, they thought it contributed to it. Their major experience with cultural unity and the interaction of science and society had been the evolution controversy of the nineteenth century. It is understandable that they considered the cultural isolation of science to be beneficial. Certainly, memories of the storms of that controversy contributed to the scientists' distaste for popular science.

This discussion of cultural dichotomy is not meant to foster the simple illusion that nonscientific culture was unified in its assumptions and interests. On the contrary, the years from 1890 to 1917 witnessed widespread cultural fragmentation and rupture with tradition. The progressive movement was evidence of the

20. Wilson, *History of the Proceedings*, pp. 5–7. Wright, *Explorer of the Universe*, p. 311.

conflict of economic and political interests between the middle class and the large corporations and unions. The middle class itself was composed of new professional groups which were specialized by education and set apart from the industrial order. Traditional Protestantism was challenged by the social gospel. Traditional economic theory was attacked by the new economics. Formalism in philosophy and law was challenged by pragmatism. The traditional arts were issuing new forms from naturalism to avant garde.[21]

On the one hand, professionalization of science was not a part of this cultural transformation. The sciences became professionally specialized and isolated in response to internal intellectual developments, such as the introduction of relativity and quantum theory into physics and statistical theory into the study of inheritance. On the other hand, this professional specialization was similar to that which many nonscientific activities were undergoing at the same time. Economists, sociologists, political scientists, architects, modern dancers, cubist painters, and muckraking journalists were all cultivating their own specialities. All of them were building separate traditions, locating the values which gave progress and meaning to their activities within the traditions themselves rather than utilizing values previously shared by all cultural activities. If scientists alienated the educated middle class when they became mathematically esoteric in physics and biology, painters alienated that class when they turned to cubism. The scientists were more fortunate than the painters, for the moment, because they seemed less endangered by this fragmentation.

Herbert Croly's discussion of the specialist in *The Promise of American Life* offers insight to the scientists' situation. Croly had been led to believe that cultural progress and unity would be achieved if all culture were based on science. He therefore con-

21. This is familiar historical territory. It is surveyed in Henry Steele Commager, *The American Mind: An Interpretation of American Thought and Character Since the 1880's* (New Haven: Yale University Press, 1950); Henry F. May, *The End of American Innocence: A Study of the First Years of Our Own Time, 1912–1917* (Chicago: Quadrangle Paperbacks, 1964); and Morton White, *Social Thought in America: The Revolt Against Formalism* (Boston: Beacon Press, 1957).

sidered science the exception to his criticisms of American culture. But since science was undergoing developments similar to
those of other activities, it would be subject to the same criticism.
Croly thought the fragmentation of society and divergent interests had produced a social crisis; by implication, they should also
produce a scientific crisis.

Croly's analysis of the specialist's situation proceeded in the
following manner. Individual excellence was defined in terms of
technical competence. If cultural unity was to be achieved, different specialists had to cooperate toward some "socially constructive" goal. They had to reach the public they were to serve. In
other words, the specialists had to become popular. Yet, "the
current American standards being what they are, how can an
individual become popular without more or less insidious and
baleful compromises?" The scientists did not think popularization was possible without such compromises. Neither were they
motivated to cooperate with other specialists. But failure to create
a public could have negative professional consequences. If there
was no consensus on the value of the scientific activity, recruit-
ment of new members into the profession would not have the
advantage of social pressure. While financial support for the
profession might be available from philanthropic sources, as it
was for science from the Carnegie Corporation and the Rockefeller Foundation, tax, subscription, and industrial grants might
be made only for reasons the scientists would consider extraneous
or damaging to their professional existence. And there was always
the vague but real problem of prestige: If a man devotes his life
to a work he considers meaningful and important, he not unnatu-
rally desires public esteem for this work.[22]

For Croly, the escape from this dilemma had to be in the popularization of specialities. But simple popularizing alone would not
be effective in converting the people to the "appreciation of
excellent special performances." The specialists had to make the
good thing look good. In other words, not popularization, but the

22. Croly, *The Promise of American Life* (1909; reprint ed., New York: E. P. Dutton,
1963), p. 442. Croly discusses the situation of an architect, not of a class of scientists; pp.
445–47.

education of the people was required to create a public. Making the good thing look good required an ideology that defined the values of the speciality, their relation to other cultural values, and the relation of knowledge in the speciality to some shared social goal. The specialists themselves constituted "the only efficient source of really formative education. In so far as a public is lacking, a public must be created."[23]

Once the scientist realized that his professional progress was not independent of other cultural developments, he would be in the dilemma of the specialist as Croly described it. Therefore, Croly predicted what the scientist would have to do once he was awakened to his cultural predicament. This awakening of the scientists was to come during and immediately following the First World War. Their experiences in the war would reveal to them the extent of their prewar isolation and the benefits accruing from shared national usefulness, assumptions, and goals. The ultimate relevance of Croly's critique of American culture was his assumption that the state was required to coordinate the special interests to serve a national goal. The government during the war did just this. The collapse of the war experience and the government's coordination of special activities at the end of the war impelled the scientists to find substitutes if they wished the accelerated scientific progress of the war to be continued in the peace. The demise of the progressive philosophy during the war and the Einstein controversy, which confused and alienated the public, added to the scientists' difficulties. These events were sufficient to induce a crisis in the scientific community.

The Impact of the First World War

The sinking of the British liner *Lusitania* in May 1915, with the loss of American and British lives, aroused George Ellery Hale's anglophilism and revived the vision of a national science in service to America. He wrote to Edwin Conklin for advice about offering the services of the National Academy of Sciences to the federal government, "assuming war, which of course may not

23. Ibid., p. 444.

come." He wanted to rush to England and France to learn the state of military science there. Conklin's temperate reaction to the events prevailed over Hale's energy. He convinced Hale that his plans were premature. Germany was not inclined to aggravate American relations. Hale's plans would be conditional.[24]

Hale's eagerness to enter science in the preparedness movement and a war effort was increased by the appointment in July 1915 of Thomas Edison to head a Naval Advisory Board for Inventions (later called the Naval Consulting Board). Physicists were not represented on the board and the fear that pure science might be left out was shared by many scientists. Arthur Webster of Clark University wrote and spoke in person to the secretary of the navy, Josephus Daniels, to ask that physicists be represented on the board. Ernest Merritt of Cornell University, president of the American Physical Society, also protested to Secretary Daniels. Merritt argued that all of invention and engineering was ultimately dependent on physics. The fundamental scientist was intellectually more creative than the engineer and could provide novel solutions to problems. A military inventions board should certainly include physicists.[25] No immediate expansion of the board, however, was forthcoming.

Despite these initial frustrations and against his own advice, Conklin took the opportunity in the next several months in conversations and letters to remind President Wilson of the National Academy of Sciences's chartered duties. He suggested, for example, the value of the academy's advice on the problem of Panama Canal landslides. President Wilson referred his suggestions to the secretary of war, but little would come of them until the sinking of the *Sussex* in late March 1916.[26]

24. The following text discusses the thought which led to the founding of the National Research Council. A fuller elaboration of Hale's private activities is available in Wright, *Explorer of the Universe*, ch. 14. Hale to Conklin, June 10, 1915; Conklin to Hale, June 17, 1915, telegram; Hale to Conklin, June 18, 1915, box 11, Hale Papers.

25. Wright, *Explorer of the Universe*, p. 286. Webster to Ernest Merritt, October 10, 1915; Merritt to Josephus Daniels, October 4, 1915, box 2, Ernest Merritt Collection, Cornell University Archives (hereafter cited as Merritt Collection).

26. Conklin to Woodrow Wilson, October 21, 1915; Conklin to Hale, October 26, 1915; Conklin to Hale, April 6, 1916; Woodrow Wilson to Conklin, October 25, 1915, box 11, Hale Papers.

This sinking and the preparedness campaign which it intensified were responsible for bringing together the national government and the National Academy of Sciences. Hale's spirit of
preparedness was virulent. He was "strongly anti-German." He
hoped that relations with Germany and Austria would be severed.
He thought America should have gone to war after the *Lusitania*
and *Arabic* sinkings and should not have allowed Germany to
laugh in her face. The American people were too easygoing, not
warlike. He hoped America would "be drawn into the war on the
side of the Allies, where we belong." He was astonished at the
innocence of the pacifists. Henry Ford and William Jennings
Bryan "ought to be imprisoned as traitors or thoroughly chloro-
formed" because they opposed intervention.[27] It cannot be concluded from these remarks that Hale would have been led to
endorse any particular kind of organization of science in America,
but they do reveal the deep emotional commitment to American
nationalism that powered Hale's effort to enlist science in any
war effort and revived the old vision of a national science.

Preparedness led to the creation of the National Research
Council as the agency of the National Academy of Sciences for
national coordination of wartime science. The council was primarily the child of Hale's imagination and efforts and, as such,
was both the fruition of his prewar vision for centralized, national
science and the foundation for postwar science. In April 1916
Hale succeeded in pressing the executive council of the National
Academy of Sciences to resolve "that in the event of a break in
diplomatic relations with any other country the academy desires
to place itself at the disposal of the Government for any service
within its scope."[28] A delegation of Hale, C. D. Walcott, Robert
Woodward, William H. Welch, and Edwin Grant Conklin from
the academy met President Wilson on April 26 to present the

27. Hale to Harry Manley Goodwin, December 12, 1915, box 2, HM 28491, and Hale
to Goodwin, February 27, 1916, box 2, HM 28492, Hale Collection. These items are
quoted by permission of the Huntington Library, San Marino, California.
28. *Report of the National Academy of Sciences for the Year 1916* (Washington, D.C.:
Government Printing Office, 1917), p. 12. See also I. Bernard Cohen, "American Physicists at War: From the First World War to 1942," *American Journal of Physics* 13 (October
1945): 333–34.

resolution. Welch, president of the academy, described the academy's chartered role and its potential for service. Hale told the president the academy could prepare an inventory of scientific personnel in the country. President Wilson made a verbal agreement that the academy should appoint a committee to proceed as the delegation suggested, adding that, in Hale's words, "he preferred to have us act with only a verbal request, as the situation with Germany is so delicate that a letter from him might be misinterpreted."[29]

Hale was appointed chairman of the new committee to plan the academy's services. The situation, he wrote to a friend, was "the greatest chance we ever had to advance research in America."[30] He soon had become convinced of a scheme for an academy-sponsored agency to coordinate all wartime research. He wrote in his diary (for June 1, 1916): "National Service Research Foundation Object: The promotion of scientific research, in the broadest and most liberal manner, for the increase of knowledge and the advancement of the national security and welfare." Hale envisioned a scheme beyond war problems and military devices. The committee chosen by Hale to help plan the research services of the academy, which included Conklin and Robert A. Millikan, completed plans in June 1916. They hesitated to bring these to President Wilson because he was campaigning for reelection on a peace platform (much to Hale's chagrin). Through the intervention of Col. Edward M. House, however, the committee's proposals were presented to Wilson, who sent a letter of approval to academy president Welch in July.[31]

29. George Ellery Hale to Evelina C. Hale, his wife, April 26, 1916, box 80, Hale Papers. See also Wright, *Explorer of the Universe*, p. 287. Helen Wright bases her account of the April 26 meeting with President Wilson on recollections of Conklin obtained in a personal interview many years later.

30. Quoted in Wright, *Explorer of the Universe*, p. 288.

31. Hale, 1916 Diary, entry for Thursday, June 1, box 91, Hale Papers. Also quoted in Wright, *Explorer of the Universe*, pp. 287–88, but in an apparently erroneous chronology of events. Hale to Col. Edward House, July 4, 1916; House to Hale, July 6, 1916; (Hale to House, July 21, 1916, referred to in the next House letter); House to Hale, July 22, 1916, box 22, Hale Papers. President Wilson's letter to William H. Welch, July 24, 1916, was reproduced in Hale, "The National Research Council," *Science*, n.s. 44 (August 25, 1916): 265.

The National Research Council, as it was called by July, was given financial support by the Engineering Foundation. The Engineering Foundation had been established for the support of engineering research by Ambrose Swasey, a machine-tool manufacturer, telescope maker, and friend of Hale. One of the foundation's vice-chairmen was Michael I. Pupin, also a friend of Hale. Foundation chairman was Gano Dunn, president of the J. G. White Engineering Corporation, a former student of Pupin at Columbia University, and another friend of Hale. In June, Pupin and Dunn asked the foundation to give financial assistance to the nascent Research Council because the National Academy had no funds for this purpose. The foundation voted its entire annual income (ten thousand dollars) to the cause, and Pupin convinced Swasey to make a personal contribution of five thousand dollars, for the year September 1916 to September 1917.[32]

In September 1916 the National Research Council officially began its career. According to the prospectus of the council, it had six duties: (1) to prepare an inventory of researchers, equipment, and current projects in the country, (2) to serve as a "clearing house" for coordination of research projects, (3) to preserve "individual freedom and initiative," (4) to cooperate with educational institutions, (5) to cooperate with research foundations, and (6) to encourage research relating to national defense and to make America independent of foreign supplies of natural resources (nitrates, for example).[33] These duties gave scientists a role in the preparedness movement and brought their professional values into agreement with the values of nationalism and patriotism.

The desire for a national science was present in other key scientists besides Hale. In Michael Pupin's views, national science was

32. Michael I. Pupin to Hale, June 2–(?), 1916, box 34, Hale Papers. An account of this was given also in Pupin, *From Immigrant to Inventor* (New York: Charles Scribner's Sons, 1923), pp. 367–68. The resolution of the Engineering Foundation, June 21, 1916, to support the National Research Council was printed in Hale, "The National Research Council," p. 266. The story of the council from April to July was repeated in Hale et al., "Preliminary Report of the Organizing Committee to the President of the Academy," *Proceedings of the National Academy of Sciences* (Washington, D.C.: National Academy of Sciences, 1916), pp. 507–10.

33. *Report of the National Academy, 1916*, p. 12. See also Cohen, "American Physicists at War," p. 334.

anchored to the basic philosophical principles of individualism and "creative coordination." He described the history of science in terms of the struggle of the individual to free himself from intellectual authority.[34] He considered individuality to be intellectual, not social, economic, or political. Seventeenth-century science was the struggle of the scientific mind to discover truths obscured by ecclesiastical authority and tradition. From the free individual mind have been derived all the sciences and material benefits which created modern civilization.

The principle of creative coordination, however, was a cosmic principle, not a historical principle like individualism, and hence the primary and more profound of the two. Life itself was the constant coordination of previously individual and independent elements. Pupin called this coordination "service" because coordination of units on one level served coordination of these units into groups on another level. Thus individual atoms were coordinated into molecules. Molecules were coordinated into organic cells. Cells were coordinated to form an organism. Similarly, independent machines were coordinated by a steam engine into a factory. The evolution of the universe was a struggle between independent units and the coordinators. It was a process of continual flux and momentary coordination.[35] The distinction of meanings of *coordinate* and *cooperate* was important. Pupin assiduously avoided the word *cooperate*. He did not think that units grouped themselves, but that they had to be forced together by an outside agency. The natural tendency of individuals was to chaos.

He extended this cosmic analysis to society without difficulty. He considered humanity's greatest problem to be the coordination of men. "The gradual solution of this problem is the evolution of social co-ordinators, which promised [sic] to lead humanity to a social cosmos." The church and the state were the major social coordinators. For the state, science was the main instrument of coordination. The activity of these three agencies would

34. Michael I. Pupin, *The New Reformation: From Physical to Spiritual Realities* (New York: Charles Scribner's Sons, 1927), pp. 3, 7–8.
35. Ibid., p. 205, and pp. 230, 237–40.

lead society toward ideal democracy. The National Research Council, Pupin said, was the first step in coordinating America's intellectuals and their societies, thus the first step toward the ideal democracy. This ideal democracy was presented by Pupin in analogy to the human body in which all cells were coordinated and pervaded by a single mind. This was the "state organism" in which the scientists were part of the "trained intellect" defining and caring for the destiny of the people.[36]

Although Pupin's social theory was similar to other modern, conservative theories, its more relevant similarity was to the social analysis of Herbert Croly. Pupin's and Croly's concerns were identical—that, left to themselves, individuals will fall into anarchy. They agreed that social harmony and democracy could not be achieved without intervention of the state using science and an intellectual elite including scientists. Croly thought only the national state could coordinate the specialists. Pupin thought only the state could relate science to society. For Pupin as for Hale, the National Research Council, with its ties to scientists, its government service, and its duty (born of the war) to coordinate scientific research, was indeed the first step to national science.

Another important scientist desiring the central coordination of science was James McKeen Cattell, the psychologist. Although he was outside Hale's circle and was not receptive to plans for increasing the power of the National Academy, Cattell did help to coordinate the activities of the American Association for the Advancement of Science with the activities of the academy's Research Council. He was secretary of the association's Committee of One Hundred on Scientific Research, which had been founded in 1913 to prepare an inventory of researchers, projects, and research support in America. During the preparedness controversy in 1916, the committee campaigned for government support of science; and when the National Research Council was founded, the committee appointed representatives to the council for those activities which they had in common. The various subcommittees of the Committee of One Hundred contained mem-

---

36. Ibid., p. 244. Pupin, *From Immigrant to Inventor*, pp. 383–87.

bers who were working also in the National Research Council, for example, Pupin and Robert Millikan on the physics subcommittee and Edwin Conklin, chairman of the biology subcommittee. Cattell was the chairman of the psychology subcommittee.[37]

Cattell strongly favored state support of scientific research, a commitment coming as much from radically democratic sentiment as from an awareness of science's needs. He was convinced that democracy was dependent on science and recommended, in a letter to William H. Welch, that a body of scientists similar to the National Academy of Sciences (of which he was a member) be made a fourth constitutional branch of the federal government for advice in lawmaking. He did not think scientific education should be left to private universities dependent on the largesse of the rich. Though it was not ideologically, economically, or politically possible before the war to achieve government support of science, the war might fulfill this goal. Cattell moved quickly, for example, to organize some scientific support for Sen. Francis G. Newlands's bill to establish engineering experiment stations at land-grant colleges. He wrote to Arthur A. Noyes (who would soon join Hale in founding the California Institute of Technology) that international competition in research should be expected after the war and should be prepared for during the war. The government had a clear duty to support research.[38]

The changes in the attitudes of the scientists connected with the establishment and work of the National Research Council

37. James McKeen Cattell, "The Committee of One Hundred on Scientific Research of the American Association for the Advancement of Science," *Science*, n.s. 45 (January 19, 1917): 57.

38. Cattell to Welch, April 26, 1913, box 10, Hale Papers. This idea was similar to Charles McCarthy's idea of the legislative reference service. Cattell, "Science, Education and Democracy" (Address delivered December 31, 1913), *James McKeen Cattell: Man of Science*, vol. 2, *Addresses and Formal Papers*, edited by A. T. Poffenberger (Lancaster, Pa.: Science Press, 1947), pp. 315–23. See also Cattell, "Scientific Research and Sigma Xi," *Science*, n.s. 41 (May 14, 1915): 731. Senate Bill 4874, introduced March 9, 1916. The bill never passed Congress. See W. R. Whitney, of General Electric, to Cattell, July 1, 1916, box 10, Hale Papers, on Cattell's activities for the Newlands bill. Cattell to Arthur A. Noyes, July 2, 1916, box 10, Hale Papers. Cattell's motive in advocating governmental support of science was precisely to free science from the philanthropic and industrial patronage that Hale sought. Cattell's important disagreements with Hale were summarized in Cattell to Hale, May 29, 1920, box 10, Hale Papers. Hale's reply is in Hale to Cattell, June 3, 1920, box 10, Hale Papers.

were discussed in a letter from Hale to Colonel House. The purpose of the letter was to secure House's aid in protecting the function of the National Research Council from usurpation by the Council of National Defense. This purpose was achieved with the appointment of the National Research Council as a department of the Council of National Defense. Hale extolled the successful cooperation and coordination of research which the Research Council had achieved between universities, private research institutions, scientific societies, and the government. They had all been brought into contact with industry. This contact would be the basis for industrial progress following the war. Hale wrote that Germany intended to wage an industrial war against the Allies at the end of the military conflict. Backed by a tradition of pure scientific research in the universities, Germany would constitute a threat to America against which an equally strong American science was a necessary defense. America was "beginning to experience some of that devotion to national service which I saw so strongly manifested in my recent visit to Europe, where all of my scientific friends . . . are actively engaged in scientific research for the government."[39] Industrial and commercial progress following the war depended on maintenance of national science. America could not go back to the prewar world.

The great example of the impact of the war in developing a new consciousness was Robert Millikan. Preeminently the commanding personality of American science and the major force in developing an ideology of science in the 1920s, he seemed the epitome of the provincial mind before the war. The son of a minister in Maquoketa, Iowa, he attended Oberlin College, where his commitment to science was made. From 1893 to 1895 he was the only graduate student at Columbia University studying for his doctorate in physics. In the latter year, he attracted the attention and sympathy of Michael Pupin, his teacher in a course on optics. When Millikan completed his doctorate, a three-hundred-dollar loan from Pupin allowed him to study at the University of Berlin. In the summer of 1896, A. A. Michelson offered Millikan an

39. Hale to House, October 31, 1916, box 22, Hale Papers.

assistantship in physics at the University of Chicago. He hocked his belongings to pay for ship's passage and returned to America to begin his career of teaching and research. He was to remain at the University of Chicago's Ryerson Laboratory until 1921, when he left to assume the head position at the new California Institute of Technology. At Chicago, Millikan was unable for a dozen years to find a research problem to which to make important professional contributions, spending this interim period in teaching, inconclusive research, and textbook composition. In 1908, he stopped writing textbooks and devoted his attention to a new research problem, the determination of the charge of an electron. Four years later, he had obtained initially satisfactory results in this experiment for which he was later to win the Nobel Prize in physics.[40]

He was elected to the National Academy of Sciences in 1915 and attended his first meeting in April 1916 to vote for Hale's resolution to offer academy services to President Wilson. He had not participated in Hale's attempted reform of the academy between 1913 and 1915. He was a member of the organizing committee which drew up the plan for the National Research Council and was later a member of the council's executive committee. In fast succession, he became chairman of the council's physics committee, the General Anti-Submarine Council, and the third vice-president of the Research Council.

Millikan's professional success and the honors to his oil-drop experiment for determining electron charge, his sudden arrival as an important scientist, the rapid rise into administrative positions of authority, the new friendships with great scientific, industrial, and governmental leaders, the new national and international problems to face—these had a great impact on the writer of textbooks and the researcher with a history of failures and dead ends.

The main scientific effort of Millikan's group of physicists was on submarine detection, one of the greater military problems of the war. The council's activities further illuminated the national

40. Details from Robert A. Millikan, *Autobiography* (New York: Prentice-Hall, 1950), pp. xi–xiii, 69.

experience of Millikan and the physicists.[41] The navy had been conducting submarine detection and antisubmarine warfare research before the National Research Council established its General Anti-Submarine Council in February 1917. The National Research Council, nonetheless, was not convinced that the navy alone could successfully develop a detection device. Besides undertaking coordination of the various groups working on the detection problem (which included the Army Coast Defense at Fort Monroe, the General Electric Company under W. R. Whitney, the Western Electric Company, and the Submarine Signaling Company at Nahant, Massachusetts), Millikan's General Anti-Submarine Council wanted to establish another research center at New London, Connecticut, comprised of university scientists. As long as university scientists were not being utilized, "there was at least ten percent of the chance [for success] that was not being utilized."[42] This plan was approved by Adm. Robert S. Griffin of Naval Engineering. In June 1917, Millikan called together a group of university scientists including Ernest Merritt of Cornell, who moved to New London for three years, A. A. Michelson, who seldom came to the station, Max Mason of Wisconsin, Harold Wilson of Rice Institute, E. F. Nichols, H. A. Bumstead, and John Zeleny of Yale. Millikan and Hale both had taken university leaves to live in Washington for full-time work for the National Research Council.

Exchange visits of scientific personnel among the Allies were an important impetus to the detection work established at New

41. There is no history specifically of scientific research in the First World War. The most detailed accounts of research are in Robert M. Yerkes, ed., *The New World of Science: Its Development During the War* (New York: Century, 1920). Aspects of this subject are touched in Cohen, "American Physicists at War"; George W. Gray, *Science at War* (New York: Harper & Brothers, 1943); and N. M. Hopkins, *The Outlook for Research and Inventions* (New York: Van Nostrand, 1919). Also Daniel J. Kevles, "Testing the Army's Intelligence: Psychologists and the Military in World War I," *Journal of American History* 55 (December 1968): 565–81; Wright, *Explorers of the Universe;* and Millikan, *Autobiography.* The papers of Ernest Merritt, Cornell University Archives, provide material for a history of the research at the New London Experimental Station from 1917 to 1920.

42. Millikan, *Autobiography,* pp. 138–39; also "Personal Recollections of the Beginnings of the New London Experimental Station," December 7, 1918, box 41, Robert A. Millikan Papers, California Institute of Technology Archives (cited hereafter as Millikan Papers).

London. As plans for the New London group were formulating in Millikan's mind, in May 1917, an English, French, and Italian scientific mission arrived, headed by Ernest Rutherford for the English. The English and French had made impressive progress in antisubmarine warfare and this was to be the basis for the American effort. This governmental service of European scientists which had so impressed Hale likewise impressed Millikan. Moreover, the experience with international scientific cooperation had great meaning for him, as contemporary documents and his autobiography testify.[43] While Hale had understood the importance of international cooperation before the war from international cooperation in astronomy and his close ties to the English scientific community, the exigencies of the war were required to give Millikan a similar understanding. But this was only reasonable. Millikan's great scientific experience before the war had been his solitary research in Ryerson Laboratory. Although Michelson worked in the same laboratory, they did not cooperate on research. Michelson was a withdrawn man, engaging little in university affairs, even eating alone. He was aloof and dignified, concerned mainly with increasing the accuracy of his instruments.[44] In the 1920s when Millikan was to deal with international questions in constructing an ideology of science, this cooperative experience with Allied scientists provided the inspiration for solution to international political problems. At the time, the exchange of visits and his impression of the national service of European scientists gave impetus to Millikan's decision to bring American university scientists to New London.

A significant event in Millikan's national experience was his taking an army commission in the summer 1917. The decision was not easily made but testified to the growing strength of this experience. He had encountered difficulty in coordinating the research of the various antisubmarine teams and resistance in the army in his attempt to coordinate researches, such as noxious

43. Millikan, "Personal Recollections"; also Millikan, *Autobiography*, pp. 152–56. See also Millikan to I. Bernard Cohen, June 30, 1942, box 36, Millikan Papers.

44. See the portrait of Michelson in Herbert Childs, *An American Genius: The Life of Ernest Orlando Lawrence* (New York: E. P. Dutton, 1968), pp. 70–71, 73.

gases research. Much of this difficulty was military resistance to civilian authority. In June 1917, Millikan was writing to his wife, still at home in Chicago, that he was to have a meeting with General Squire of the Army Signal Corps to determine whether to take a commission as major.[45] By July it had become imperative to do so if he were to accomplish the National Research Council's work.

This decision involved a genuine transformation of his role in the war effort. Before taking his commission, he had been a university scientist with ties to the smaller world of personal research. Taking the commission severed him from his past and generated a more profound commitment to national service. The meaning of the commission was revealed in his letters to his wife Greta, especially in a July 25, 1917, letter. He told her that his family duties were primary, governmental duties were secondary. Nevertheless, he was becoming deeply implicated in the war effort. Unless he got out at that moment, getting out could be impossible. Michelson had written asking him to return to Chicago for the September term. He was tempted:

> But when I have been thinking this way and then get hold of a paper such as appeared this morning, and realize that we are up against the most terrific struggle in history and that the issue of that struggle depends on us every instinct that I have says go to it absolutely regardless of the costs and play your part and if I'm alone I shouldn't hesitate a minute to cut loose from the University entirely if necessary and come back to it when the war is over if they want me then.[46]

Millikan undoubtedly felt deeply about the war news. This breathtaking explanation of his decision to take a commission ran on, without stops, as if it had been rehearsed often and that to have interrupted it would have diminished its force or weakened its credibility.

45. Robert A. Millikan to Greta Millikan, June 26, 1917, and June 29, 1917, box 53, Millikan Papers. The official background of Millikan's commission is discussed in "The Work of the National Research Council," *Science*, n.s. 46 (August 3, 1917): 99–100.
46. Robert A. Millikan to Greta Millikan, July 25, 1917, box 53, Millikan Papers.

Millikan's national experience was paralleled by that of other members of the National Research Council's General Anti-Submarine Council. Ernest Merritt, the Cornell University physicist, joined the navy and moved to New London. At the end of the war, Merritt sent a memorandum to his naval commander urging that the New London Experimental Station not be closed. Merritt himself continued to work on submarine warfare problems at New London until 1920. At that time he sent a memorandum to Secretary of the Navy Daniels again urging peacetime maintenance of an experimental station with a skeleton staff and a full staff on immediate call. He did not think the kind of peacetime research he had in mind could be accomplished in the regular naval testing and development program because engineers could not strike off in new research directions. Future submarine research might require the aid of biologists or chemists, for instance. Continual preparedness would prevent the next war from being fought like the last.[47] It would also, of course, alter the professional role of advisory scientists by tying them to governmental interests.

Harold Wilson, an English physicist attached to Rice Institute who also moved to New London, while not sure that the war would push peacetime pure science into new scientific problems, was sure that the war had changed the sociology of American science. The war had awakened the scientists and the nation to the practical value of scientific knowledge, and this recognition should be reflected in increased emphasis on scientific education and in a diversion of scientific activity from problems of pure science to commercial and applied science.[48] Such problems were far more tied to broader cultural and national interests than was prewar pure science.

Were it not that Millikan and the other scientists were leaders of science, it would be trivial to observe that their war effort made a strong impression on them. It is not trivial because this

47. Merritt to Captain J. F. Defrees, U.S.N., November 9, 1918, box 3; Merritt to Josephus Daniels, January 21, 1920, box 2, Merritt Collection.
48. Harold A. Wilson, "Science After the War," Rice Institute Pamphlets 6 (1919): 319–21.

national experience led them to create a new basis for science in America. Hale was unusual in having a full vision of national science before the war, but the war intervened in the professional lives of Millikan, Webster, Merritt, Wilson, and perhaps Pupin, to give them also a commitment to national science. Reflecting on his wartime activities in his autobiography, Millikan wrote that he had joined the National Research Council with the two objectives of assisting the preparedness campaign, then the war effort, and "to help as well as I could in laying the foundations for the best possible development of American science."[49]

The experiences of these scientists during the war and their newly generated commitments were formally summarized and rationalized in the revealing document, "Suggestions for the International Organization of Science and Research," submitted to the Interallied Conference on International Scientific Organizations in London, June 1918. The suggestions were made during a meeting of the executive council of the National Academy of Sciences at the Cosmos Club in Washington, April 21, 1918.[50] Present at the meeting were Hale, Conklin, Michelson, C. D. Walcott, A. A. Noyes, Whitman Cross, and W. H. Howell. The suggestions were placed under four headings: "Immediate Importance of International Organization to Meet War's Needs," "New Factors in the Organization of Research Emphasized by Recent Experience," "Establishing of National Research Organizations," and "Establishment of an International Research Organization." The second discussion is most pertinent, but the other discussions place it in context.

The war made obvious to all scientists that the special sciences would gain from international cooperation. It was suggested that an international organization was needed to promote the general interests of science. Public appreciation of the "national impor-

---

49. Millikan, *Autobiography*, p. 168.

50. Arthur L. Day, home secretary, "Suggestions for the International Organization of Science and Research: Submitted by the Council of the National Academy of Sciences," council minutes, April 21, 1918, box 53, Hale Papers. The "Suggestions" were published in the "Third Annual Report of the National Research Council," *Report of the National Academy of Sciences for the Year 1918* (Washington, D.C.: Government Printing Office, 1919), pp. 52–54.

tance" of science had greatly increased and the time was ripe for a vigorous international movement on behalf of scientific research. The scientists did not seem aware of the latent contradiction in this discussion between nationally important science and the international promotion of science. Certainly the unity of national and international war goals, at least in propaganda, masked this contradiction. For these scientists there was also disillusionment at Versailles.

The members of the council thought that other countries ought to establish national academies and national research councils similar to those of America. This would insure in all countries the direct contact of science with the military and economic requirements of the government, educational and industrial needs in scientific research, and the problems of the public welfare. In America, science was to lessen the burden of taxation by reducing the costs and prices of manufactured goods and to increase agricultural productivity. To accomplish these objectives all scientists would have to be "strongly united." Independent researches would have to be coordinated. The interests of pure science and applied science would have to be federated to preserve "a just balance" between their needs and their usefulness. The council of the National Academy was making one of its strongest appeals for national science.

Important as these suggestions were, they all sprang from the more important lessons which the war experience had provided the scientists. The first lesson was "the essential solidarity of scientific research in all of its aspects." The scientists found during the war no difficulty in assuming administrative duties and in undertaking project research (as contrasted to research in which the scientist was free to pursue whatever natural phenomenon occurred). This experience was to be important in the 1920s when the scientists sought to convince the American people that scientific research united many activities of the government and the economy. The war experience could also be, as for Millikan, personal evidence that the popular image of the scientist as an impractical and esoteric theorist was incorrect.

The second lesson followed naturally from the first: the necessity for cooperation between academic, industrial, and military researchers.

The third lesson was a statement of the national experience: "the increased sense of the obligation to the state of all men of science, and their strengthened desire to render their researches of the widest possible value." The activities of Hale and Millikan in organizing American science, and of men like Merritt and the others in lesser roles, surely evidenced this profound impact of the national experience. The leaders of science agreed that American science would not return to the prewar situation.

The final four lessons of the war were affirmations of the scientists' intention to project the wartime organization of science into the peace. These were the advantages accruing from cooperation of specialists in different fields and contact with military and industrial problems; the need to strengthen the nation through research, that is, preparedness; the recognition by industrialists of their obligation to support pure science; and "the importance of taking immediate advantage of the present opportunity to promote the interests of science and research."[51]

Preparing for Peacetime Science

The scientists attempted to transfer the new organization of wartime science into the postwar era in three major ways: by the permanent establishment of the National Research Council, by the proposal for national research laboratories in physics and chemistry, and by the activities of the Division of States Relations of the National Research Council.

An executive order from President Wilson was required to transform the National Research Council from a wartime to a permanent agency. It had been Hale's and Millikan's intention from the council's inception to make it permanent. Several months after America's entry into the war, Hale revealed his long-range plans for the Research Council in a statement prepared for the Carnegie Corporation. In "The National Academy

51. The preceding discussion was based on the manuscript copy of the "Suggestions."

of Sciences and the National Research Council as Factors in National Progress," Hale sought to justify a peacetime council. By reference to his abortive efforts at reform of the academy before the war, he sought continuity which made the council seem less of a war measure. He also argued the importance of the council for American culture and progress beyond its importance for professional physics.

The ability of the National Academy to respond to sudden national needs was proved by its creation of the Research Council. This response should have indicated the effectiveness with which the academy would use an endowment. If it had a home building for itself and the council, there would be a "more effective utilization for national ends of existing research agencies."

The war had increased scientific activity in other nations. This was not the least factor pressing for continuation of wartime national science and preventing a return to prewar scientific provincialism. "Governments that have felt the pressure of military and industrial necessity are now vieing [sic] with one another in promoting scientific research." After the war, international research would have an important influence on national industrial development. Hale, too, expressed this latent contradiction between international cooperation and national competition.

What had to be done was clear, Hale concluded. There must be cultivation of a close relationship with the state and industry; coordination of all governmental, educational, and privately endowed research agencies; and the promotion of cooperative international research. All the nation's research should be brought into "a single federation." This was the function of the National Research Council.[52]

52. It was not clear at the time what legal necessity existed for an executive order to make the council permanent, because the council was private and not an agency of the government. The official status of the council was as a subcommittee of the National Academy. Apparently, Hale and the others thought that an executive order would raise the prestige of the council, aid in convincing the National Academy to designate it a permanent subcommittee, and help to raise its funds. This discussion is based on George Ellery Hale, "The National Academy of Sciences and the National Research Council as Factors in National Progress: Statement Prepared for the Trustees of the Carnegie Corporation," August 30, 1917, box 41, Millikan Papers. The purpose of this statement was to secure funds for a National Academy of Sciences building.

After presenting his case to the Carnegie Corporation and while waiting for its decision on his request for endowment, Hale and the Research Council members sought to convince President Wilson to issue an executive order permanently establishing the Research Council. Hale wrote to Wilson on March 26, 1918, asking for the order and submitting supporting documents. One document was Wilson's July 24, 1916, letter to William H. Welch in which the president wrote that he would ask bureau heads to cooperate with the council. Hale's request, however, was confusing with regard to what specifically he wanted the executive order to say. Wilson revealed this confusion when he asked Hale, "Is it your idea that the Research Council should be given *authority* to coordinate the work now being done by the regularly constituted scientific agencies of the Government, or is it merely that the purposes of the Council as it is already operating should be more explicitly set forth?"[53] Wilson doubted both his right and the wisdom to give the council such authority.

Hale denied that authority was desired. The council wanted only official recognition to increase its influence and therefore ability to do what it was doing. Primarily such an executive order would assure the council "continuance . . . under future administrations." But Hale's denial of desire for authority seemed belied by the original draft for the executive order which he had suggested and sent to Wilson. That draft contained the provision that one of the council's functions would be "to serve as a correlating and centralizing agency for the research work of the Government." Along with the denial, Hale suggested as a substitute for this phrase the following: "to promote co-operation among the scientific and technical bureaus of the Government." This cooperation could be achieved by presidential appointment of bureau chiefs to the council.[54]

---

53. Hale to Wilson, March 26, 1918; Wilson to Hale, April 19, 1918, box 41, Millikan Papers.

54. Hale to Wilson, April 22, 1918, box 41, Millikan Papers. Despite the loss of the centralizing function for the council's governmental relations, Hale was pleased with the executive order as promulgated. See George Ellery Hale to Evelina Hale, May 10, 1918, box 80, Hale Papers.

The executive order of President Wilson establishing the council was promulgated on May 11, 1918.[55]

The desire to centralize science in America through the National Research Council was further revealed in a confidential, anonymous memorandum that circulated in May 1919 among the members of the executive council of the Research Council. Although describing the wartime origin of the council, the memorandum was written in the present tense. This excerpt contains the bluntest statement of the council's purpose.

> The Academy organized the National Research Council, which is intended to be a federation of all the important scientific and technical agencies of the country, military and civil, governmental, industrial and academic, with a view to stimulating the growth of science and its application to industry, and particularly with a view to the co-ordinating of research agencies for the sake of enabling the United States, in spite of its democratic, individualistic organization, to bend its energies effectively toward a common purpose.[56]

C. D. Walcott and Hale also described the council's purpose in terms of federation of science in a February 1919 letter to the Carnegie Corporation asking for a five-million-dollar endowment for the academy. They explained that governmental support for the council from the president's Emergency Fund would soon be exhausted and probably not renewed. If the council were to operate on a permanent basis, private assistance would be required. In support of this letter, they submitted a statement, "The Purpose and Needs of the National Research Council," which outlined the federal character of the council.[57] The divisions of the council, for example, physics, chemistry, engineering, govern-

55. The order was reprinted in the "Third Annual Report of the National Research Council," pp. 25–26. See also Daniel J. Kevles, "George Ellery Hale, the First World War, and the Advancement of Science in America," *Isis* 59 (1968): 427–37.

56. "Confidential: The Origin and Purpose of the National Research Council," box 41, Millikan Papers. The manuscript bears the date May 21, 1919, although it seems to have been written some time before then. The author was probably Hale.

57. Hale and Charles D. Walcott to James Bertram, secretary, Carnegie Corporation, February 1, 1919, box 67, Hale Papers.

mental relations, had on their boards representatives of local scientific societies or government bureaus and commissions. Chairmen of the divisions had a year's tenure and were paid for full-time employment. This assured continuity of the council's activities.

On March 28, 1919, the Carnegie Corporation approved a grant of five million dollars to the academy and council for the long-sought building and to provide an endowment for an operating income.[58]

The permanent establishment of the National Research Council by President Wilson and the endowment of the National Academy and Research Council by the Carnegie Corporation insured the continuation of national science. But these developments did not solve the problem of the extent to which peacetime science would be centralized. The controversy over the extent of centralization is instructive because it reveals how far the scientists had moved from their prewar views. There was little desire to return to the diffused organization of private science of 1913–1914. This dispute over the centralization of science at the national level versus regional centralization was misleadingly and inconsistently called "centralization versus federation."

The controversy was initiated by a proposal of the Rockefeller Foundation for national laboratories for research in physics and chemistry. This proposal indicated a second way in which the scientists tried to carry the wartime organization of science into the peace. In February 1918, George Vincent, president of the Rockefeller Foundation, wrote to Robert Millikan for his response to the establishment of "an institution . . . devoted to pure research, unhampered by obligations to teach and uninfluenced by commercial consideration" which would provide "leadership in American progress in the physical sciences." Millikan showed the letter to Hale. The two immediately disagreed about the proposal. As he wrote to Vincent, Millikan thought that the establishment of a central institute for research similar to the Kaiser

58. Copy of the "Resolutions Adopted by the Board of Trustees of Carnegie Corporation of New York at Meeting of March 28th, 1919," in box 67, Hale Papers.

Wilhelm Institute at Dahlem, for example, was an obvious remedy for the American deficiency in pure science, but not the best.
The foundation's money could best be used by building up the
research facilities of universities around the country.[59]

Hale and Simon Flexner of the Rockefeller Foundation supported the scheme for the central laboratory. For Hale, it was the
opportunity to realize his prewar plans for National Academy
laboratories. For the foundation, it was an effort to do for physics
and chemistry what the Rockefeller Institute for Medical
Research had done for medicine. Hale and Flexner thought that
such a central institute would benefit rather than compete with
universities associated with it.[60]

Millikan relates in his autobiography that he and Hale con-
vened a group of sixteen physicists and chemists to vote on the
question of a central laboratory versus six around the country.
The vote went seven to nine in Millikan's favor. This obviously
was not a decisive vote. In an attempt to compromise and secure
the Rockefeller funds, Millikan proposed a scheme for postdoc-
toral fellowships in physics and chemistry to be taken to any one
of six designated university laboratories. This proposal coincided
with Hale's own longstanding scheme for science fellowships.[61]
Although this was not mentioned in his autobiography, Millikan
also was opposed to having the National Research Council award
these fellowships.[62]

Hale was reluctant to oppose Millikan's opinions on the laboratory and fellowship schemes. He did not doubt that Millikan was
as concerned as he was to build a national science and disagreed
only over organization. Hale wanted the National Research
Council to award the fellowships, but, if necessary to keep the

59. Vincent to Millikan, February 5, 1918 (Millikan reprinted this letter in his *Autobiography*, pp. 180–81); Millikan to Vincent, [no date available because the manuscript is
torn], box 5, Millikan Papers.

60. The present-day Brookhaven National Laboratories is probably closest to the idea
envisioned by the foundation and by Hale.

61. Millikan, *Autobiography*, pp. 183–84. See Wright, *Explorer of the Universe*, pp.
305–06, for an account of Hale's fellowship scheme.

62. Millikan to Hale, June 15, 1918, box 29, Hale Papers.

scheme alive, he would agree that up to six designated universities should make the awards.[63]

A compromise proposal appeared shortly after President Wilson issued the executive order establishing the Research Council.[64] It contained an elaborate rationalization for scientific research beginning with the familiar argument that after the war "the sharpest industrial competition ever known" would occur. The nation with the best scientific establishment would triumph. Furthermore, industrial research in science would be financed by industries out of self-interest. Pure scientific research had to be increased because industrial (that is, product-oriented) research depended on it.

This particular argument led to familiar contradictions. The National Academy's goal of international scientific cooperation was contradicted by the argument that the postwar period would witness industrial competition based on science. The second contradiction was fully exposed in this proposal. The argument that pure research ought to be supported by industries was contradicted by the following statement. "Self-interest, once fully aroused, will amply provide for industrial investigations; but this is not true of fundamental scientific research on which industrial progress depends." Because industries would not support pure science, the proposal argued, the Rockefeller Foundation should. Both contradictions sprang from the basic situation of public science in private and competitive settings. No nation would support open, cooperative research whose results might be to its disadvantage in international competition. Likewise, no industry would support open, pure research if it would result in a competitive disadvantage. The war effort had masked these difficulties by uniting all national endeavors for a single goal. These contradictions would not be masked in the 1920s, however, and together with other contradictions they would doom the attempt to formu-

63. Hale to Millikan, June 21, 1918, ibid.

64. National Research Council, "Proposal for the Endowment of Research in Physics and Chemistry," [no date, but necessarily written between May 11, 1918, and October 3, 1918], box 5, Millikan Papers.

late an internally consistent ideology of science and the appeal for industrial grants based on the ideology.

After making its rationalization for pure science, the proposal outlined the compromise scheme for support of pure science. Three research centers, each affiliated with a suitable educational institution, were proposed. Each center would have a permanent staff and a rotating staff composed of National Research fellows. There were also to be traveling fellowships for research abroad. The national fellowships would have a tenure of two to five years. The centers, fellowships, and visits by foreign scientists would be "administered and co-ordinated by a central agency, the National Research Council." The council would also undertake cooperative research. Each cooperating university was to contribute two million dollars. The Rockefeller Foundation was to contribute three hundred thousand dollars a year for ten years to support the fellowships, foreign visitors, and administrative costs.

The proposal stated that this scheme was better than that of the central laboratory because universities would benefit from cooperation rather than suffer from competition. It was also better than merely dispersing Rockefeller Foundation funds in small grants because it preserved the "numerous advantages of concentration and association."

Before the myriad demands on universities' funds in the 1920s, this compromise scheme was doubtlessly doomed. It must also have seemed too centralized for Millikan though this can only be conjectured. But the fellowship plan was accepted by the Rockefeller Foundation.[65] Fellowship applicants had to designate at what institution they proposed to do research, allowing the council fellowship board to insure their going to the best laboratories and indirectly allowing it to build up certain institutions, such as the California Institute of Technology in the 1920s and the phys-

65. The Rockefeller Foundation officially endorsed the plan on April 9, 1919. See the letter from Edwin R. Embree, secretary of the foundation, to Hale, April 10, 1919, reprinted in "Extracts from the Minutes of the Meeting of the Executive Council [of the National Research Council]," *Proceedings of the National Academy of Sciences* (Washington, D.C.: National Academy of Sciences, 1919), p. 308. Public announcement of the fellowships was made on March 29, 1919, and was reprinted in ibid., p. 315.

ics department of the University of California at Berkeley in the 1930s. Without question, the National Research fellowships were of enormous importance in building up the physical sciences in America in these decades.[66]

Though the research centers scheme was never realized, the proposal was argued into the 1920s. There remained the inhibiting division between the proponents of a national laboratory, including Hale, A. A. Noyes, Gilbert N. Lewis, the great chemist at Berkeley, and Simon Flexner, and their opponents, including Vernon Kellogg, John Campbell Merriam, president of the Carnegie Institution, and Millikan.[67] In 1920, the National Research Council made another proposal for the research center to the Rockefeller Foundation. The opposition to a central laboratory was so adamant, however, that the proposal contained two schemes, one for the central laboratory and one for strengthening selected university laboratories. The Rockefeller Foundation was to take its pick.[68]

This presentation made explicit the basic agreement on the need to reorganize physical science in America. The disagreement was about the character of the reorganization and the distribution of authority. The introductory statement to the proposal cited the shortage of researchers and teachers created by postwar

66. Two holders of National Research fellowships in physics were Leonard Loeb, son of the biologist Jacques Loeb and Millikan's student at the University of Chicago before the war, and Ernest Lawrence, inventor of the cyclotron. The story of Loeb particularly reveals the benefits to American science conferred by the fellowship plan. It can be followed in the correspondence between Jacques Loeb and Robert Millikan, 1913 to 1919, box 9, Jacques Loeb Papers, Manuscript Division, Library of Congress, and in the National Research Council files, box 5, Millikan Papers. The story of Lawrence is told in Childs, *An American Genius*.

67. This was the lineup remembered by Millikan in 1923 when the national laboratory idea was momentarily revived. Millikan to Kellogg, November 14, 1923, box 5, Millikan Papers. In 1921, Millikan backed away from his opposition to the national laboratory scheme to the extent that he would accept two national laboratories, if they were affiliated with universities. Millikan wanted the California Institute of Technology, which he was about to head, to be one of these. The other could be the University of Chicago, which he was about to leave. See Millikan to Hale, May 12, 1921, box 29, Hale Papers.

68. A. A. Noyes to Millikan, May 7, 1920; "Plan for the Promotion of Research in Physics and Chemistry: Prepared by the Research Fellowship Board of the National Research Council, May 1920," box 5, Millikan Papers. The following discussion is based on the "Plan."

increased college enrollments. It made no reference to the expected industrial competition which had previously been an important argument for the support of basic science. The disagreement between the central institute and regional laboratories schemes was emphasized in this statement regarding the central institute, that "the *trusteeship* of the funds of the Institute and the ultimate *legal authority* shall be vested in the National Academy of Sciences." The policies and appointments of the institute were to be established by a board of governors appointed by the president of the National Academy. The purpose of the alternative plan of Millikan was "greatly to strengthen the graduate departments of physics and chemistry at a number of selected universities so as to create a group of important centers of research widely distributed throughout the country." A small annual sum (that is, twenty thousand dollars) would be granted to each center which would match at least half the amount. All authority would be held by the university at which the center was located.

The main advantage of the national institute would be in the "more complete experimental equipment" it would have. It would be free of institutional rivalries which prevent cooperation. It did not inhibit any university from enlarging its own research facilities. The main advantage of the alternative plan was that it would immediately double research facilities and stimulate research in all regions. To the criticism that it would promote provincial competitiveness, its proponents argued that provincialism would be eliminated by gathering researchers at different centers in different years.

Neither of these plans was carried out, though the history of university research in America has come close to fulfilling the regional laboratories scheme. Some universities have had nationally famous departments which acted as centers of research. But these have risen and fallen with endowments, success in faculty raiding, and student-teacher ratios.[69]

69. A historical summary of the shifting prominence of research centers for all scientific fields is available in Stephen Sargent Visher, *Scientists Starred, 1903–1943, in "American Men of Science"*: *A Study of Collegiate and Doctoral Training, Birthplace, Distribution,*

The National Research Council's attempts to centralize science in America were not exhausted by the research or fellowship plans. The Division of States Relations, under the administration of Albert L. Barrows, executive secretary, also sought to coordinate research projects around the country. Barrows's views illustrate the impact of the wartime national experience on the country's scientists.

The Division of States Relations was established in a general reorganization of the National Research Council in February 1919. The division's object was to coordinate the state research councils which had been created by states' councils of defense during the war. The division would have had an obvious role in coordinating research of different states, in facilitating the exchange of personnel and information, and in the allocation of research labor. It was hoped, for example, that cooperative research could be undertaken in areas like sanitation and livestock conservation.[70]

Barrows's letters to the state councils still functioning after the war summarized the National Research Council's rationalization of coordinated research. The letter to Joseph Hansen, chairman of the Idaho Scientific Research Committee, was typical. After asking for information about activities during the war, Barrows explained the National Research Council's purpose. The service of scientists during the war had demonstrated the dependence of industry, education, and public welfare on the progress of scientific knowledge. The council desired to continue the war effort's "earnest attention to scientific affairs." Adjustment to peace depended on scientific research. "The National Research Council

Backgrounds, and Developmental Influences (Baltimore: Johns Hopkins Press, 1947), ch. 5.

70. "Extracts of Minutes of Joint Meetings of the Executive Board of the National Research Council with the National Academy of Sciences," *Proceedings of the National Academy of Sciences* (1919), pp. 181–87. "Extracts from the Minutes of the Meeting of the Executive Board," in ibid., p. 304. Albert L. Barrows to John Campbell Merriam, chairman of the Executive Council, August 21, 1919, box 221, John Campbell Merriam Papers, Library of Congress, Manuscript Division (cited hereafter as Merriam Papers). A more complete statement of the purpose of the Division of States Relations is available in its "Preliminary Announcement," [January or February 1920], box 226, Merriam Papers.

is now considering whether such state committees, perhaps with certain modifications, may not constitute a useful permanent agency for applying science to the good of the people and at the same time stimulating scientific progress so that still further benefit may be had from it." Writing to A. H. Hirst, president of the American Association of State Highway Officials, Barrows described the state research councils as the local representatives of the National Research Council.[71]

Barrows personally wanted the council to be more agressive. He thought it was too eager to return to the "academic view of progress in science." He meant that the council should have been less concerned with the problems of science in the universities than the problems of science in the nation. "There has been but little expression as yet of definitely relating the Council to the needs of the nation on the same basis of patriotism as that which it served during the war." He considered national, agricultural, and conservation problems to have been on the same level as the war's problems. He admitted that the council was moving in the direction of national science but felt it was not moving fast enough. "I have wondered if it might not be in order to express a policy of applying its energies upon two or three carefully de-fined national issues . . . with somewhat the same force with which the Council declared its support of the preparations for war." If science were to be so organized, public appreciation and support would guarantee progress in the smaller world of pro-fessional science.[72]

Albert Barrows's impatience with the National Research Coun-cil emphasized the direction in which the council was moving. It provided a measure of the distance which the scientific commu-nity had moved from its prewar views and aspirations. And it summarized the impact of the national experience of the war. Scientific leaders realized that American progress in science required the end of professional provincialism and prevented a

71. Barrows to Joseph Hansen, September 8, 1919; Barrows to Hirst, September 10, 1919, box 221, Merriam Papers.
72. Albert L. Barrows to David P. Barrows, November 28, 1919, box 226, Merriam Papers.

return to the smaller prewar world. These leaders realized also the enormous benefits accruing to science from unification with the rest of American culture. Almost as if Herbert Croly's *Promise of American Life* had been a prophecy for science, to end the alienation between specialists and laymen and to relate the values of science to those of American culture were to be prime occupations of scientific leaders in the 1920s.

The collapse of the war effort ended the quasi-legal authority of the National Research Council to coordinate American science and the abnormal unification of American culture. Yet coordination of science and unification of culture were prerequisites for the national science which scientists now wanted. If these wartime gains were to be preserved, a new consensus about the place of science in American civilization had to be substituted for the war effort. This was the crisis of science and democracy and the task of the ideology of science in the 1920s.

(3) SCIENCE SERVICE

The Revival of Popular Science

The demand of the preparedness movement for national and cultural unity provoked the scientists to reexamine their isolation from the American public. The demise of the tradition of popularization, which previously had been acceptable, now became intolerable because it prevented the scientists from explaining to the laymen the contributions they could make to the war effort and from eliciting the public's sympathy and support. A number of scientists immediately responded to this situation by proposing a new journal of popular science. A journal could be the ideological organ for national science.

The first effort to establish a journal came in 1916, at the same time the National Research Council was being founded. During July of that year, George Ellery Hale was attempting to generate enthusiasm among the scientists for a popular magazine to be sponsored by the National Academy. Hale did not meet an entirely favorable response. With a hint of criticism and impatience, Millikan wrote to his wife that Hale was "the most restless flea on the American continent—more things are eating him." Millikan opposed the proposed journal because he thought its objective could be accomplished as well by existing magazines.[1] Hale's effort for the journal among the scientists of the National Academy was paralleled by the effort of William J. Humphreys

1. Robert A. Millikan to Greta Millikan, July 19, 1916, box 53, Robert A. Millikan Papers, California Institute of Technology Archives (cited hereafter as Millikan Papers).

among the members of the American Association for the Advancement of Science. At Humphreys's instigation, on September 30, the American Association's committee on policy discussed the sponsorship of a popular journal. A subcommittee composed of Humphreys (as chairman), S. W. Stratton, chief of the Bureau of Standards, and Hale was appointed to investigate and report on the journal scheme. The policy committee thought the proposed journal could stimulate public interest in science and publicize the importance of research to the nation. Humphreys himself apparently considered the promotion of the journal a patriotic duty because of the importance of science for preparedness. He also hoped the American Association and National Academy could cooperate in the sponsorship of the journal.[2] A third effort to establish a popular science journal was undertaken at this time by the American Chemical Society. Despite some overtures for cooperation between the two national scientific societies and the chemical society, it was decided by the national societies not to include the chemists in their venture.[3]

The report of the subcommittee of Humphreys, Stratton, and Hale was presented to the Committee on Policy at its December 1916 meeting in New York. The report asserted that national preparedness demanded a wide diffusion of scientific knowledge: "The vital importance of scientific knowledge to national preparedness . . . is so abundantly obvious that it requires no argument. Nor is it any less obvious that the value of knowledge increases in proportion to its diffusion." The report suggested that the attractiveness of the journal to scientists could be enhanced by initially restricting its subscription list to the memberships of the National Academy and the American Association.

2. Hale was appointed without his prior knowledge. L. O. Howard, secretary, "The Committee on Policy of the American Association for the Advancement of Science," *Science*, n.s. 44 (October 13, 1916): 526–27. Edward L. Nichols (Cornell University), chairman of the committee on policy, to Hale, October 5, 1916; Humphreys to Hale, October 7, 1916, box 75, George Ellery Hale Papers, The Carnegie Institution and the California Institute of Technology, Pasadena, California (cited hereafter as Hale Papers).

3. Arthur D. Little, writing for the American Chemical Society, to L. O. Howard, secretary of the American Association, October 18, 1916; Hale to Humphreys, November 11, 1916; Humphreys to Hale, November 13, 1916, box 75, Hale Papers.

The report was accepted and a new subcommittee formed to plan the financing and publication of the proposed journal.[4]

Opposition to the proposed magazine soon appeared. Hale and the group of scientists who founded the National Research Council were not successful in convincing the scientific community that, as the subcommittee report assumed, the effectiveness of scientific knowledge increased with its diffusion. E. B. Wilson of Harvard, for example, thought that any new journal would merely duplicate *Scientific Monthly*, edited by Cattell, and *Popular Science Monthly*, edited by Walter Kaempffert. Wilson also had " 'a rather poor opinion of scientific men for running a popular journal.' "[5] There was just the slightest intimation in Wilson's tone that it was beneath the dignity of scientists to explain their work to laymen.

The important organized opposition to the popular journal scheme came from the National Academy scientists at the University of Chicago, including Millikan, A. A. Michelson, Thomas C. Chamberlin, a geologist, and J. M. Coulter, a botanist. A statement of February 13, 1917, signed by all the Chicago members, presented their objections. They felt that the journal's supporters confused the importance of scientific knowledge for national preparedness with the importance of an additional popular magazine. The Chicago scientists thought that the importance of science for preparedness depended on the productiveness of scientists, not on the diffusion of scientific knowledge. Productivity and diffusion of knowledge were neither identical nor mutually dependent. They did not think National Academy members should divert their energies from scientific research to popularize science. They opposed calling in nonscientific literary writers to run the popular journal. If any organization was to sponsor such a journal, the American Association for the Advancement of Sci-

4. Hale actually had no hand in the composition of the report, but subscribed to its recommendations. Humphreys to Hale, December 20, 1916, box 75, Hale Papers. Under this scheme, the journal's first subscription list could have been just over eleven thousand. William J. Humphreys, S. W. Stratton, and George E. Hale, "Report of the Journal Committee," [December 1916], box 75, Hale Papers. Arthur L. Day to Hale, January 16, 1917, ibid.

5. Quoted in Hale to Arthur L. Day, January 10, 1917, ibid.

ence alone was the obvious candidate. But they doubted that the American Association needed to add to existing magazines and doubted, moreover, that it was capable of managing a good magazine.[6] During the war, scientists like Millikan were to lose their doubts concerning the importance of popular science. The war convinced them that national scientific productiveness could not exist without wide popular support.

Though the journal's advocates attempted to counter the criticism of the Chicago group, the opposition of the latter, together with the emergency presented by America's entry into combat, successfully prevented the National Academy and the American Association from publishing the proposed journal.[7]

Following the war, the drive to establish the popular science magazine was revived. The National Research Council was being reorganized to coordinate peacetime scientific research on a national level. The unification of scientific research and national goals provided by the war was in danger of being dissolved in the peace. The scientists turned to the lay public to discover both the national goals that would justify national coordination of science and the support for the maintenance of national science. The proposed popular journal was to be a major instrument for building the public consensus for this national science. The change in attitudes of some scientists toward popularization was striking. Robert Millikan by 1920 was emerging as the leading popular spokesman of the scientists. He was one of the founders of Science Service and one of its trustees. Henry Fairfield Osborn, who in 1913 had opposed Hale's reforms, including popularization, endorsed the planning for a popular journal in 1919. Hale, of course, encouraged the change in attitudes, assuring the scientists that the National Research Council's postwar objectives of coordinating and stimulating research would be well served by a journal which could increase public appreciation of science and

6. This statement of National Academy scientists at the University of Chicago, February 13, 1917, was enclosed with the letter from Arthur L. Day to Hale, February 17, 1917, ibid.

7. The reply to the criticisms of the Chicago group is in Humphreys to Day, February 24, 1917; discussion of the postponement of the effort to establish the journal is given in Hale to Day, March 5, 1917, ibid.

"better comprehension" among industrialists of the necessity for basic research.[8]

Although the leaders of science now wanted a popular science journal, it proved difficult to finance. The National Academy at this time was gathering funds for the academy building, the research fellowships, and the operating budget of the Research Council. The proposed journal had to take a low priority for the moment. Vernon Kellogg, permanent secretary of the Research Council, was asked by the council to study the problem of the journal's financing. Because of the difficulties of locating support, Kellogg moved toward a plan for making the journal commercially successful. In September 1920, he talked with Merle Thorpe, founder of Nation's Business, about the feasibility of a self-supporting popular science journal under the auspices of the National Academy and the American Association.[9] Before the popular science journal scheme could reach fruition, however, Science Service was established, usurping both the proposed journal's function and its affiliations with the National Academy, the National Research Council, and the American Association for the Advancement of Science.

Science Service was the direct outgrowth of the American Society for the Dissemination of Science, founded in 1919 by the newspaper magnate, E. W. Scripps, and his close friend, William Emerson Ritter. The transformation of the American Society for the Dissemination of Science into Science Service demonstrated the scientists' determination to build a consensus for national science.

E. W. Scripps had founded thirty-odd newspapers and the United Press, and had amassed a huge fortune in doing so. His papers had a reputation for radicalism because of their occasional endorsements of the single tax and socialism. Scripps himself was an ardent champion of social democracy, always prescribing more

8. Hale to Henry Fairfield Osborn, May 17, 1919, ibid.

9. John Campbell Merriam to Hale, June 6, 1919, box 28, Hale Papers, reviews the unsuccessful attempts to finance the proposed journal. Kellogg to Hale, September 21, 1920; Kellogg to Hale, October 2, 1920; Hale to Kellogg, October 11, 1920; Kellogg to Hale, October 20, 1920, box 75, Hale Papers.

democratization to cure democracy's ills. In collaboration with William Emerson Ritter, a biologist at the University of California who had great admiration for him, Scripps founded the Scripps Institution for Biological Research in La Jolla, California, in 1903, the Foundation for Population Research at Miami University in 1921, and, of course, Science Service in 1921.[10]

Scripps's desire to popularize science sprang from his deep anxiety for democracy. He thought that democracy conceived as "the right of the many to have the same influence on the whole body politic as the exceptionally few and able" was socialism. He opposed this. But he did favor democracy if this was conceived as the right of the community "to select the man best qualified to govern the community in that department of its affairs where a man of his qualifications is required."[11] A democratic community should be one in which every person knew his own interests and sought his proper place. Democracy, therefore, depended on self-knowledge and education. Scripps was not convinced of the value of institutional education. He thought that critical self-education obtained as a person worked through life was more useful.[12] Scripps's anxiety for democracy and his concern for education easily combined with a concern for science, which his friend Ritter gave to him. Drawing on his experience as a journalist, in 1919 Scripps formulated a plan for the popularization of science. An adult citizenry, constantly educating itself in the advances of science, was necessary to maintain democracy in a scientific civilization.

Scripps believed in the efficacy of science as an instrument of human welfare. In his later years, he came to believe that industrial and economic forces were driving America toward a catastrophe which only science could avert. He believed quite

10. Gilson Gardner, *Lusty Scripps: The Life of E. W. Scripps* (New York: Vanguard Press, 1932), p. 222. There is no scholarly biography of Scripps. Gardner's biography contains the essential biographical information and a well-drawn portrait of Scripps's character. Oliver Knight, ed., *I Protest: Selected Disquisitions of E. W. Scripps* (Madison: University of Wisconsin Press, 1966), pp. 725–29.

11. E. W. Scripps, "Advancing Democracy," March 27, 1909, in Knight, *I Protest*, pp. 388–89.

12. Gardner, *Lusty Scripps*, pp. 168–72.

literally that science was to save man. Science Service was the embodiment of this faith.[13]

The American Society for the Dissemination of Science, as Scripps had first envisioned it in 1919, was to be composed of scientists and researchers with a lay journalist as editor. The group was to prepare scientific news stories for distribution to newspapers. This plan was clearly based on that of the Newspaper Enterprise Association, organized by Scripps to supply features to newspapers unable to sponsor them alone. Scripps supposed that the Newspaper Enterprise Association would be an important client of the popular science agency. The American Society for the Dissemination of Science was not to be a profit-making venture, beyond the needs of maintaining its service. Scripps would give thirty thousand dollars annually to underwrite the venture and expected to give a total endowment of five hundred thousand dollars.[14]

Scripps apparently asked Ritter to discuss this scheme with scientists. Writing to various scientists, Ritter inadvertently encountered those in the National Academy, National Research Council, and the American Association for the Advancement of Science who were working on the plan for a popular science journal. When this latter group learned of Scripps and Ritter's plan, they immediately sought to absorb it into their own plans. As William J. Humphreys wrote after Ritter had approached him: How "opportune" that "when the scientific men themselves had just reached that stage in their own plans for writing for the public," Scripps's offer to finance a writing service came along! Humphreys thought that the proposed popular science journal of the national scientific societies would be an obvious client for Scripps's science writing service. But even more, Humphreys suggested, perhaps Scripps's writing service could come under the "general supervision" of the National Research Council, since the Research Council was the best source of scientific infor-

13. William Emerson Ritter, "Relation of E. W. Scripps to Science," *Science*, n.s. 65 (March 25, 1927): 291–92.
14. This discussion is based on Scripps's disquisition, "The American Society for the Dissemination of Science," dictated on March 5, 1919, box 75, Hale Papers.

mation for popular articles. Humphreys told Ritter that he had discussed Ritter's letter and scheme with C. D. Walcott, president of the National Academy; J. R. Angell, president of the Research Council; and L. O. Howard, president of the American Association for the Advancement of Science. They concurred that Scripps's proposal was desirable. Humphreys also sent copies of Ritter's letter to Hale, D. T. MacDougal of the Desert Laboratory, Arthur A. Noyes, John C. Merriam, and W. W. Campbell.[15]

The response to the Scripps-Ritter proposal indicated how far the majority of scientists had moved away from their prewar indifference or hostility to popularization. The agreement of the national leaders with Ritter's letter indicated the temper of the scientific community. Ritter wrote that he had interviewed over three hundred scientists and journalists with regard to the proposal. He found that the "great majority . . . are greatly impressed with the importance of some such effort as that proposed by Mr. Scripps."[16]

Acting on the suggestion of Humphreys, Ritter wrote to C. D. Walcott asking that representatives of the National Academy, Research Council, and American Association for the Advancement of Science be appointed to attend a meeting at Scripps's estate in San Diego in March 1920 to discuss Scripps's plan.[17] Those present at the meeting included D. T. MacDougal for the American Association for the Advancement of Science, Hale for the National Academy, and Millikan for the National Research Council. Scripps, Ritter, and Robert P. Scripps (E. W. Scripps's son) represented the American Society for the Dissemination of Science. At the first meeting, March 17, E. W. Scripps read his paper, "Notes on the New Society," outlining his proposal. Scripps added to the original proposal by suggesting the initiation of a popular science lecture circuit and endorsed support for a popular science journal. He denied any desire to have any influence on the politics of the new writing service. The following day, this group dissolved the American Society for the

15. Humphreys to Ritter, January 15, 1920 (a copy), ibid.
16. Ritter to Hale, March 1, 1920, ibid. I have no breakdown of Ritter's statistics.
17. Ritter to Walcott, March 1, 1920 (a copy), ibid.

Dissemination of Science and formally reorganized as an independent agency, Science News Service (which would be given the final name Science Service), with affiliations with the National Academy, Research Council, and the American Association for the Advancement of Science. It was decided also to include a journalist on the governing board.[18]

The selection of an editor for Science Service was a most difficult task. The editor would have to be knowledgeable in science as well as a good writer. He would have to have managerial talents. But primarily he would have to have the concern for national science which was held by the group which now controlled the new service. Few had these qualifications. Those with the qualifications, like James McKeen Cattell and Walter Kaempffert, were already editors. After a long period of deliberation, the Service in 1920 formally offered the editorship to Edwin E. Slosson, associate editor of the *Independent*, formerly a chemist, and author of popular science articles and books. This offer placed Slosson in an awkward position. He had also been asked by Vernon Kellogg to be the editor of the proposed popular science journal which the national scientific organizations still considered sponsoring.[19]

The personal views of Slosson on popularization were finally to decide the fate of the proposed popular science journal and the character of the Science Service. He thought that a popular science journal of the quality of *National Geographic*, for example, would help the American Association for the Advancement of Science serve its nonprofessional interests and fulfill its original purpose which was the promotion of useful knowledge. More than this, however, such a magazine was necessary for the educated class of persons who wished to be informed of the continu-

18. Secretary's notes, "Informal Meeting Held at Miramar, California," March 17, 1920; E. W. Scripps, "Notes on the New Society," delivered March 17, 1920; secretary's notes, "Meeting of the American Society for the Dissemination of Science," March 18, 1920, Miramar, California, box 5, Millikan Papers.

19. Mary E. Scott, secretary, "Minutes of the Meeting of the Science News Service Held at Miramar, July 7, 1920," box 75, Hale Papers. At the meeting, Millikan resigned as representative of the National Research Council and was replaced by Arthur A. Noyes. Vernon Kellogg to George Ellery Hale, July 27, 1920, box 25, Hale Papers.

ing progress of science, for the scientific specialist who wished to know about developments outside his field, and for the self-educated. It was possible to compose accurate and comprehensible articles about science for an audience which had at least the equivalent of a high school education. "The aim should always be to show that science does not consist of fixed formulae or a code of laws handed down from heaven on tablets of stone like the Ten Commandments, but the growing thoughts of living men groping after greater knowledge." This scientific search for knowledge ramified throughout modern society. And Slosson favored an editorial policy of exposé and muckraking to distinguish the pseudoscience from the genuine science, and of ridicule for antiquated beliefs.[20]

But Slosson was ultimately doubtful of his ability for business management and was reticent to accept a magazine editorship, if that would be one of his duties. He suggested, therefore, that the efforts of the proposed popular science journal be combined with the Science Service.[21] The independent journal was never established. In December 1920, Slosson accepted the editorship of Science Service; he wrote to Ritter that popularization was the primary interest of his life. In January 1921, in the rooms of the National Research Council, the work of Science Service was begun.[22]

The Message in Edwin Slosson's Popular Science

Edwin E. Slosson was ideally suited to be editor of the scientists' new agency for popularization. Indeed, his life seemed to be a preparation for this post, which he took only ten years before his death. A native Kansan, he had attended the University of Kansas where he received both his bachelor's and master's

20. Edwin E. Slosson, "Note on the American Association for the Advancement of Science," enclosed with Slosson to Ritter, August 17, 1920 (a copy), box 75, Hale Papers. Slosson, "Suggestions in regard to a popular science periodical," enclosed with ibid.
21. Slosson to Ritter, August 17, 1920, ibid.
22. Slosson to Ritter, November 20, 1920, ibid.; Kellogg to Hale, January 8, 1921, box 25, ibid.

degrees in science. His classmates included William Allen White and Vernon Kellogg. He taught chemistry at the University of Wyoming for twelve years, attending summer sessions at the University of Chicago to earn his doctorate in organic chemistry, which he received in 1902, magna cum laude.

As a result of popular articles he wrote for the *Independent* while at Wyoming, he was called to New York in 1903 to be an editor of that magazine. He remained with the *Independent* until he went to Science Service in Washington, D.C., in 1921. The move to New York was good for Slosson. It thrust him into the center of American intellectual life. Immediately he loved urban culture, the turmoil of traffic as well as the opera. He gained a large number of friends including John Spargo, the socialist (also on the *Independent* staff); John Dewey; James Harvey Robinson; James T. Shotwell, the Columbia University historian; Corra Harris, the Southern belletrist; David Lambuth and J. P. Turner, biologists at Woods Hole, where he often passed summer vacations; Thomas A. Edison; L. H. Baekeland, the inventor of the plastic Bakelite; and Wayne B. Wheeler, leader of the Anti-Saloon League. As an *Independent* editor he interviewed the great intellectual and artistic figures of the day, including H. G. Wells (who became his friend); Maurice Maeterlinck, the Belgian poet; Henri Bergson; and G. K. Chesterton.

His literary productiveness was prodigious. In his lifetime, Slosson published twenty technical articles based on professional research in chemistry, eighty pamphlets, eighteen books, and about two thousand signed articles, essays, and editorials. He wrote also for the *Independent* three to four thousand words of news summary a week, and for a long time he wrote all the magazine's unsigned foreign news. He pioneered in writing movie reviews.[23] In all of this, there were over four hundred popular science articles.

Slosson's concern for popular science was essentially that of the

23. The biographical details are from Preston William Slosson, "Editor's Foreword," and "E. E. Slosson, Pioneer," in Edwin E. Slosson, *A Number of Things* (New York: Harcourt, Brace, 1930), pp. ix, 3–33; Edwin E. Slosson, *Sermons of a Chemist* (New York: Harcourt, Brace, 1925), p. vi.

scientists. He was convinced that the American public was basically hostile, or, in their best moods, indifferent to physical science. This hostility was not only misplaced, but dangerous. The world war had demonstrated that the preservation of American democracy depended on the researches of basic science and the public's appreciation of these researches.[24] The goal of Science Service was to establish a public consensus on the importance of national science. This concern of the scientists and this objective of the news service found their paradigmatic formulation in a long letter from William Emerson Ritter to E. W. Scripps. This letter was written shortly after Science Service was established.

> I do not believe it is too much to say [Ritter wrote] that there are few really open, alert minded scientific men in the country who do not recognize, more or less clearly, that the continued progress if not the actual existence of our nation are now in the balance. Probably no scientist anticipates such a collapse for us as has lately befallen several European nations. Rather they see in the sum total of present conditions and tendencies evidence that unless far reaching modifications in our national life are brought about, the peak of our curve of growth and prosperity is reached, and from now on we shall be sliding down, perhaps very gradually, the other slope of the curve.
>
> I am also sure that a goodly number of scientists are convinced that a far wider dissemination among the people than now exists of the results of scientific investigation and of the methods and mental attitude of science is indespensible [sic] if such national down-sliding is to be averted.
>
> And finally, we know for a certainty that a considerable number of men of science believe that something similar to Science Service is absolutely necessary as one means for accomplishing the results desired.[25]

24. Slosson's general point of view was presented in "Talk to Trustees of Science Service at the Meeting of June 17, 1921," especially sections headed "Hostility Toward Science," and "The De-personalization of Science," box 194, John Campbell Merriam Papers, Manuscript Division, Library of Congress (cited hereafter as Merriam Papers).
25. Ritter to Scripps, May 13, 1921, ibid.

Social upheaval in Europe and economic recession in America threatened traditional American beliefs and progress. Ritter's conviction that science should strengthen these values and guarantee progress was shared by other important scientists, like Robert Millikan. But Ritter's letter was more than an expression of the American conservative reaction to World War I. It was a statement of the ideological purpose of popularization. Popularization had to persuade the public to accept the scientists' values. This consensus would provide the cultural unity necessary for national science and, as well, preserve America's traditional character and beliefs.

Edwin Slosson was certain to direct Science Service to its ideological goal. His own personality embodied the cultural unity he hoped to promote in America. His intellectual experience with the *Independent* forced him to establish relationships between the values of science and those, for example, of pragmatic liberalism, engineering, and muckraking journalism. Slosson's ideas led directly to the ideology of science.

To accomplish its ideological task, popular science had to demonstrate a fundamental proposition: *that scientific ideas conceived within the framework of theories about the nature of the physical world were the primary causes of cultural change.* Stated less precisely, this proposition asserted that pure science was the basis for progress. The fundamental proposition was at once the scientist's correction of the layman's erroneous impression that the scientist's work was not socially beneficial and also the basis for the scientific progressive's faith that modern civilization would improve its living standard, increase democracy, and raise the moral level of men—in Ritter's words, that national down-sliding could be averted.

The fundamental proposition was not widely held during the 1920s. There was in this decade, as the philosopher Morris R. Cohen said, an "insurgence against reason." From many sides, the efficacy and power of *idea* were denied. The real cause of cultural change and the real motivations of men were not deliberated ideas, so the attack went, but were historical forces, interests, libido or stimuli, or techniques. The deliberated idea was only self-deluding rationalization, hiding what really caused change and actions. The disillusionment with the Versailles settlements, for example, discredited rational purposes. The goals which supposedly guided America's entry into the war turned out to be only propaganda masking the real economic forces and national interest at work. At the time, some intellectuals claimed that their ideas had led the American entry into the war, but as Randolph Bourne wrote in his famous essay "War and the Intellectuals," their ideas were "little more than a description and justification of what [was] going on." After the war, Walter Lippmann doubted that democracy was the best form of government because he did not think that the average citizen could gain correct ideas of situations on which to make a decision to vote.[26]

The prevailing theories of personality development, Freudianism and Watson's behaviorism, also discredited the power of the idea. The popular interpretation of Freudianism suggested that all personal behavior was caused by some condition of the libido or sexual instinct. An idea such as the "good end," which a person might think guided his actions, was only self-delusion. John B. Watson's behaviorism likewise denigrated the efficacy of the idea by asserting that all behavior was only the response, however complicated, to antecedent and external stimuli. An idea was only the internally stored stimulus.

Another attack on the power of ideas was made by the engineering profession. Since the progressive period, the engineer had argued that the major cultural changes were initiated by introduction of new technologies. The enthusiasm of the progressive period and the 1920s for engineering led many persons to believe that cultural progress was achieved by increasing the efficiency and order of technology, rather than by thinking up new ideas.

For Slosson, the scientists, and scientific progressives like John

26. Morris Raphael Cohen, *Reason and Nature: An Essay on the Meaning of Scientific Method* (1931; reprint ed., London: Collier-Macmillan, 1964), ch.1. The substance of this chapter was published in "The Insurgence Against Reason," *Journal of Philosophy* 22 (February and March 1925): 113–26; 141–50. Randolph S. Bourne, "War and the Intellectuals," in *War and the Intellectuals: Essays by Randolph S. Bourne, 1915–1919*, ed. Carl Resek (New York: Harper & Row, Torchbooks, 1964), p. 12. Walter Lippmann, *Public Opinion* (1922; reprint ed., New York: Macmillan, 1961), pt. I.

Dewey, James Harvey Robinson, and Charles A. Beard, ideas had a primary and creative role in man's interaction with his environment. To prove the fundamental proposition and to deflect attacks against it, Slosson offered in his popular science three categories of evidence: philosophical, sociological, and historical. The philosophical proof attempted to demonstrate a necessary relationship between idea, technique, and environment. The sociological proof traced the products of modern industrial and urban culture to origins in scientific ideas which were conceived without regard to their possible social effects. The historical proof attempted to demonstrate that scientific ideas were the causes of historical change as well as cultural change. All of these demonstrations were circumscribed by the character of popular science which prohibited intensive reasoning. But this absence of logical rigor in the presentations of evidence does not mean that what Slosson was attempting to prove was trivial. To demonstrate how our ideas affect reality is the basic task of an ideology.

Slosson addressed himself at greatest length to the three aspects of demonstrating the fundamental proposition in "Science Remaking the World." In a preface to "Wonder-Working Gasolene," Slosson prepared the reader for the theme of the series of articles. Although a new scientific idea was often scoffed or laughed at, he said, the idea "soon sinks into the mass of accumulated knowledge where it works as an unconscious ferment. In the course of a generation or two, perhaps, the world wakes up to the fact that its mind has changed." To ascribe the action of ideas to an unconscious ferment would convince no one of the efficaciousness of ideas. Therefore, Slosson turned to discussions of commonplace items such as gasoline, refrigeration, and photography to delineate precisely how ideas work.[27]

Slosson argued that a natural phenomenon did not produce cultural change without the intervention of man. This interven-

---

27. Edwin E. Slosson, "Science Remaking the World," *World's Work* 45 (November 1922-March 1923). The individual articles were "I. Wonder-Working Gasolene," (November 1922), pp. 39–50; "II. Cold Almost as Useful as Heat," (December 1922), pp. 162–75; "III. Coal Tar as a World Power," (January 1923), pp. 255–65; "IV. The Influence of Photography on Modern Life," (February 1923), pp. 399–416; "V. The Influence of Sugar-Power in History," (March 1923), pp. 495–508.

tion had to be, first, a conception of the phenomenon in an idea. This first intervention was not the application of a technique to the phenomenon. Thus, the natural properties of gases did not produce cultural change in themselves. The unrestricted movement of gaseous molecules did not, for example, produce motion. But when man gained scientific knowledge of what would happen to the gaseous molecules when placed in a container and the size of the container changed, he had the essential conception of the piston. The idea of the piston, together with other ideas, such as the translation of reciprocal motion into rotatory motion, yielded the idea of the engine. Slosson summarized the point of this example, that the intervention of the ideas into natural phenomena was necessary to produce change, in the words that "force directed by intelligence produces progress."[28]

What Slosson was saying about the relationships between idea, technique, and environment was clear: to conceive of the world was to conceive of its possible alterations. Grasping a natural phenomenon in an idea made possible the manipulation of nature. The idea could be changed, dissected, analyzed, and recombined, until the idea of the natural phenomenon yielded an alteration that an available technique could wrest out of the phenomenon itself. Ideas did not imitate nature; they conquered nature. Slosson illustrated this by the invention of artificial cold. Few phenomena could be as useless to man as natural cold. Nevertheless, whether as an ice cave or a refrigerator, artificial cold benefitted man, by preserving foods from decay, for example. Artificial cold was not an imitation of nature, any more than purpose and benefit were imitations of anarchy and uselessness. A subtler illustration of this thesis was provided by infrared photography. Without sensory aids, man cannot perceive the heat landscape. But the extension of the idea of photography of visible light to the photography of invisible radiation provided men with new ways to perceive and manipulate the visual world.[29]

28. Slosson, "Science Remaking the World, I," p. 44.
29. This notion was explored in an elliptical manner in Slosson's diatribe against primitivism, "Back to Nature? Never! Forward to the Machine," *Independent* 101 (January 3, 1920): 5. Slosson, "Science Remaking the World, II," pp. 162–75; "Science Remaking the World, IV," p. 402.

Slosson's philosophical interpretation of the idea, carefully underlying the popularizations of the scientific advances above, was primarily indebted to John Dewey's pragmatism and instrumentalism. Dewey's opposition to the observer theory of perception, which stated that the idea was merely a passive copy of nature, and his instrumental theory of cognition were thoroughly known and subscribed to by Slosson. Dewey maintained that all thinking was a process of problem-solving, and that all knowledge—the end result of thinking—was judgment. Thinking did not occur unless problems were presented in experience. Ideas were the plans of action to solve the problems. Judgments were ideas confirmed by successful carrying out of the plans of action. Dewey said that thinking "appears as the dominant trait of a situation when there is something seriously the matter, some trouble, due to active discordance, dissentiency, conflict among the factors of a prior non-intellectual experience."[30] We do not need to enter into the epistemological aspects of Dewey's philosophy, namely, his resolution of the classical percept-object problem, to understand Slosson's indebtedness to Dewey's instrumental concept of idea.

When the kill of yesterday's hunt has spoiled in the hot sun of today, or when the legs of the table that we have just built do not touch the floor together, a process of reflection is initiated. Dewey described four stages in this process. First, doubt is realized. Experience has conflicts within itself, or with our expectations, and we doubt that the situation is as it should be. Second, we rummage among our collection of fixed ideas describing such a situation and rules for meeting doubtful situations to locate one appropriate to this occasion. When, third, the situation is not resolved by the old idea or rule, we alter the idea to fit the situation. If the altered idea does not solve the problem, we seek new

30. Edwin E. Slosson, "Pragmatism," *Independent* 62 (February 21, 1907): 422–25, reprinted in *A Number of Things*, pp. 135–43. A rancorous discussion of the pragmatic theory of perception was provided by the Introduction to *Intelligence and the Modern World: John Dewey's Philosophy*, ed. Joseph Ratner (New York: Modern Library, 1939), pp. 3–241. John Dewey, *Essays in Experimental Logic* (Chicago: University of Chicago Press, 1916), p. 11.

facts and widen our experience. Seeking new facts is the fourth, and scientific, stage of reflection. With the new facts, we fashion still newer ideas for resolving the doubt. In this process, the "idea" appears as a rule for resolving the problem. Old ideas, designated and fixed by words, simply are traditionally successful rules for grappling with the issue. When they fail, new ideas must be conceived. In this way, the idea is a symbolic substitute for the problematic situation itself. We can manipulate the idea to derive an alternative, which if successfully applied to the problem by a technology, would be confirmed as valid or true knowledge. Ideas, therefore, are "tentative, dubious but experimental, anticipations of an object. They are 'subjective' (i.e., individualistic) surrogates of public, cosmic things, which may be so manipulated and elaborated as to terminate in public things."[31] Accepting Dewey's theory, Slosson could argue that ideas do not imitate nature, but conquer nature.

Slosson was able to draw even more than the instrumental theory of ideas from Dewey's pragmatism. Dewey provided also the reason why ideas were always causes of social progress. Ideas, in Dewey's view, arose out of social situations and found their terminus in social situations. The problems which initiated thinking were problems by reference to human needs. If the word *needs* gives to Dewey's philosophy an anthropomorphism he would have denied, one can say, instead, that some situations were problems because of man's presence in nature. Thus, the idea, if it was a correct idea, was always the solution of social problems and yielded social progress. Thinking and the formulation of ideas never took place isolated from the social situation. They never occurred as pure intellection. Dewey put it this way: The "essential feature" of the formulation of ideas was "control of the environment in behalf of human progress and well-being, the effort at control being stimulated by the needs, the defects, the troubles, which accrue when the environment coerces and suppresses man or when man endeavors in ignorance to override

31. Dewey, *Essays in Experimental Logic*, p. 228. Dewey's discussions of the process of thinking and the character of ideas are in chs. 6, 7, and 8.

the environment."[32] This was the philosophical reason why scientific ideas always were the primary causes of cultural change and why ideas were not self-deluding rationalizations.

The sociological demonstration that scientific ideas must cause cultural change provided most of the content of popular science. This emphasis was not simply, as Slosson occasionally said, to provide human interest for otherwise uninteresting information. Nor was this abundant description of telephones, coal-tar dyes, radios, and all the accruements of modern scientific civilization merely to prove the utilitarian benefits of science. Of course, both human interest and the marvels of science helped to build the consensus on the importance of science. But the real purpose of the sociological description was the proof of the fundamental philosophical proposition about ideas and cultural change. Of course, sociological proof of philosophical propositions is, at best, a dubious matter. It would have been easier to prove sociologically that techniques change technologies, which, in turn, change technologically derived economics and institutions. But this would have established the primary value of engineering, not of science. The engineers did, indeed, argue that cultural changes were caused by an independent technological tradition which only occasionally used the scientific method of discovery. The engineers were interested in the organization and construction of the environment, not the discovery of new facts or the enlargement of the environment. Slosson and the scientists had to show the primacy of scientific ideas and the dependency of technology on them in the achievement of progress.[33]

If it had been obvious to the public that modern culture was the product of scientific ideas, writers of popular science could have merely recited the scientific wonders of the day because the public would have known the point of such a recitation. Almost always, however, the marvels of the day were related to scientific ideas by some argument which minimized the independence of

32. Ibid., p. 22.
33. See, for example, Slosson's comments in his discussion, "The De-personalization of Science," in "Talk to Trustees of Science Service," June 17, 1921, box 194, Merriam Papers.

technical tradition, the trial-and-error method of the inventor, and the pressures of industrial economics. The argument emphasized the unity of the scientific method and scientific ideas (technologists, therefore, could not have used the former without using the latter) and the independence of the creative scientist from the industrial revolution. In this manner, the scientists' ideology was distinguished from the ideology of the engineers.

The best illustrations of the sociological proof of the fundamental proposition were contained in Slosson's *Creative Chemistry* and the articles "Science Remaking the World." The growth of cities in this century, promotion of international commercial unity, increase in the democratic spirit, change in courtship patterns, and numerous secondary and tertiary effects such as the growth of a huge rubber industry and increased tax revenues were attributed to the mass-produced automobile. The automobile was made possible by development of the compact, gasoline, internal-combustion engine. The engine, in turn, had been derived from a knowledge of petroleum, thermodynamics, and gases. And this knowledge was gained by scientists working free from any knowledge of the twentieth-century applications of their ideas.[34] This argument was presented in a pseudodeductive manner, as if the tertiary economic effects were only the logical implications of Boyle's and Gay-Lussac's gas laws. A more pointed example of this argument concerned coal tar. "Coal-Tar Colors" began with a description of the incidental production of coal tars from the industrial distillation of coal gas and the manufacture of coke. The coal tar was dumped as waste. The industrialists and industrial engineers had no idea of the potential use of coal tars, "being for the most part ordinary persons and not born chemists." It required a chemist, attempting to prepare artificial quinine from coal tar in the laboratory, to notice that coal tar when mixed with alcohol yielded a purple solution. This was William Henry Perkin and the discovery was of synthetic dyes in 1856. The reason that a scientist and not an engineer conceived the idea of artificial dyes from this purple solution was that sci-

34. Slosson, "Science Remaking the World, I," pp. 42–50.

ence was a method of discovery and problem-solving, whereas industrial and technological research primarily were not. Slosson discussed the complex chemistry of the coal tar (that is, organic chemistry) in this work. It was the knowledge of organic chemistry which the scientist had, and which was necessary to formulate new ideas about coal tar, that the engineer did not have.[35] The engineer may have designed the technique to manufacture the synthetic dyes, but this clearly came after the scientist's ideas.

The connotation of inevitability of sociological progress deriving from scientific discovery, as in the illustration above, came in part, unfortunately, from the logical fallacies of the presentation—its pseudodeductive and post hoc, ergo propter hoc character. But the connotation of inevitability came mainly from the implicit faith in scientific progress, revealed in such statements as "the revolutions of science are achieved without the use of arms or argument," and "the scientist does not have to persuade or convince anybody; he either has discovered something or he has not."[36] The inevitability of scientific progress and the consequent inevitability of cultural progress were considered evidence for the pragmatic theory that an idea always has social consequences.

Although the concept of progress in the seventeenth and eighteenth centuries came out of the scientific revolution, the belief in progress in the nineteenth and twentieth centuries had come to support, in a circular manner, the belief that scientific ideas always have social consequences. The treacherous logic of this situation was fully revealed in Slosson's essay, "Is There a Law of Human Progress?" Henry Adams and Alfred Korzybski each thought that he had discovered the law of progress. Adams thought eras of human change followed the law of squares. Korzybski thought (in *The Manhood of Humanity: The Science and Art of Human Engineering*, 1921) that change was in geometrical ratio to the elapsed time. Slosson considered both "laws" ficti-

35. Slosson, *Creative Chemistry: Descriptive of Recent Achievements in the Chemical Industries* (New York: Century, 1919), pp. 60, 64–69, 72, 74. Slosson discussed this theme again in "Science Remaking the World, III," pp. 256, 262–63.
36. Slosson, "Science Remaking the World, V," p. 508.

tious, but he did think that they were based on a genuine principle, "that ideas interact if they get together." As long as ideas interact, progress occurs. Slosson did not think there was a *law* of progress, because *law* implied that progress occurred regardless of condition. Progress was at least conditional on science. Science was the activity of discovering facts, formulating new ideas, and synthesizing them. Of course, this was all a tautology, saying only that if there was progress, there was progress. Thus, for example, Slosson would explain cultural lag as the failure of ideas to get together. The obviousness of the logical situation eliminates the need for an extended examination of it.[37]

Slosson was led by this logic to maintain that even the most abstruse mathematical ideas had utilitarian and social consequences. No more than any other scientific ideas were those of mathematics pure intellection or self-delusion. "Just as soon as a mathematician in the seclusion of his study writes down a new formula, the paper is snatched from him by the hand of some mechanic who carries it off to his shop and sets it to work."[38]

Slosson's popular science maintained that scientific ideas were the primary causes of sociological change and of progress. Were scientific ideas also the primary agents of historical change? This was the third aspect of Slosson's demonstration of the fundamental proposition.

The question whether scientific ideas were the cause of sociological change is not identical to the question whether scientific ideas were the cause of historical change. Sociological change need be only quantitative, for example, increasing the size of cities, decreasing the extent of manual labor. For this reason, Slosson's illustrations usually concerned scientifically induced technology, such as the electrical technology based on Faraday's discoveries and ideas. Progressive and cultural change, however, imply historical change. And historical change is qualitative. That is, historical epochs cannot be deduced from one another and

37. Slosson, "Is There a Law of Human Progress?" *Independent* 108 (February 25, 1922): 185–86.
38. Slosson, "Science in Daily Life," in *Snapshots of Science* (New York: Century, 1928), p. 93.

future history cannot be prophesied. If Slosson was to demonstrate conclusively that scientific ideas were the chief causes of cultural change—the fundamental proposition—he would have to prove that scientific ideas were the chief cause of qualitative, historical change as well as of quantitative, sociological change.

It was not obvious, however, that scientific ideas were the cause of qualitative, historical change. The attempt to prove that they were led immediately to a dilemma, stemming from the identification of progress with qualitative, historical change, and outlined as follows. If Slosson failed to prove that scientific ideas were the causes of qualitative, historical change, then he could not prove that progress constituted such change. More importantly, if Slosson did successfully prove that scientific ideas caused qualitative, historical change, then he had also proved that there had been no such change since the scientific revolution. This latter entailment is made clear when the meaning of *qualitative, historical change* is examined. One historical epoch is distinguished from others by its chief characteristic or causal agency. When progress is defined as the cultural movement from one historical epoch to another, as Slosson wanted, then progress must be marked by new causal agencies. Thus passage from the Middle Ages to the modern era was caused by the scientific revolution. Since Slosson would never have relinquished his major tenet, that the modern—present—era is distinguished by science, the conclusion must be that there has been no progress since the scientific revolution, that is, that culture has not moved out of the epoch initiated by that revolution. But since Slosson also insisted that progress was now occurring, another conclusion is that progress was not necessarily qualitative, historical change—only quantitative, sociological change. The implications of this dilemma strengthened the technologists' argument that technology caused progress, the notion Slosson originally hoped to subvert.

Needless to say, Slosson approached this intellectual dilemma with caution. One of his standard pronouncements was that scientific revolutions resulted in qualitative, historical change, while political revolutions did not. The phrase *political revolutions* certainly connoted qualitative change. To have substituted scientific revolutions for political revolutions as the basic determinant of historical change should, in a strict comparison, have made scientific revolutions bearers of qualitative change. Slosson wrote, "Political revolutions merely change the form of government or the name of the party in power. Scientific revolutions really turn the world over, and it never settles back into its former place." Socialism and capitalism were "merely by-products of the laboratory." He repeated this thesis in "Science Remaking the World." He even reiterated the nineteenth-century myth that science erased political divisions. He argued that coal-tar derivatives had increased democracy by providing mass luxuries, that photography had done likewise by providing mass art, and that central heating had broken up the family structure by allowing the members to disperse to their separate, heated rooms.[39]

Slosson's "Chemistry and the Past" gave the strongest statement that scientific ideas caused historical change. He thought that chemistry initiated historical change in two senses: first, that basic chemical processes, such as a change in the oxygen content of the ocean, initiated new historical eras; second, that man's conquest of nature through scientific (in this case, chemical) ideas initiated new eras. The first category of causation was illustrated in Slosson's explanation for the decline of the Hanseatic League in the late fourteenth and early fifteenth centuries. He suggested that the rise of the League cities, based on the Baltic Sea fisheries, caused an increase in the carbon dioxide content of the Baltic. This increased carbon dioxide content drove the herring into the North Sea. Because the North Sea was fished by the English and Dutch fleets, the fortunes of England and the Netherlands rose, while those of the Hanse declined.[40] This explanation was, of course, pure speculation and not even historically accurate. The trade monopoly of the League had been broken by

39. Slosson, "Science and Journalism," *Independent* 74 (April 24, 1913): 913, 914; "Science Remaking the World, I," pp. 39, 45–46; ibid., "III," p. 255; ibid., "IV," pp. 400, 416; and "The Disintegration of the Family Circle," in *Short Talks on Science* (New York: Century, 1930), pp. 114–18.

40. Slosson, "Chemistry and the Past, with Interesting Predictions as to the Future," *Century* 120 (January 1930): 32.

the English, Dutch, and Scandinavians in the late fourteenth century before the herring had begun to spawn in the North Sea about 1417. More important, however, chemical processes accompany all environmental and historical events. Which particular reaction accompanies what particular historical event depends on historical factors, such as political decisions.

Slosson was on firmer ground when he suggested that man's chemical knowledge initiated new historical eras.[41] The transition from the Bronze Age to the Iron Age, like the transition from the bow and arrow to gunpowder, was made possible by chemical knowledge. Again, however, Slosson's failure to address himself directly to the question of causation meant ambiguity with regard to whether such a historical transition was caused by the increase in chemical knowledge or merely made possible by that knowledge.

Ingenious as were these illustrations of qualitative, historical change, they lacked the defining feature of such change: the emergence of a new mind. James Harvey Robinson knew this in the 1920s when he had called for a new mind appropriate to the new scientific age. On occasion, Slosson recognized this also. In one essay, he wrote, "The great revolutions are those that change not only customs and industries but the mind of man, that alter his outlook on the world at large, his philosophy of life."[42]

For Slosson, the scientists, and the scientific progressives, the great revolution which differentiated modern history from previous history was the scientific revolution. Slosson, Millikan, Hale, Pupin, Robinson, and Beard shared this belief that seventeenth-century science had cut man away from the forces of history and nature and given him the ability to shape his own destiny. This belief, of course, has been one of the central beliefs of Western man since the eighteenth-century Enlightenment.[43] Although the belief suffered a temporary eclipse in the second half of the nineteenth century because of the dominance of social

41. Ibid., p. 29.
42. Slosson, "Science Remaking the World, IV," p. 402.
43. On this, see Peter Gay, *The Enlightenment—An Interpretation: The Rise of Modern Paganism* (New York: Alfred A. Knopf, 1966).

Darwinism, it had been fully revived as a central tenet of modern liberalism before the world war.

Slosson characterized modern history as the "creative period" of man's evolution. Because of his own experience as a chemist and from a shortsighted view of the chemical revolution of the nineteenth century, Slosson tended to regard chemistry as the science that gave man his creative capabilities. The chemist's ability to synthesize substances gave man an artificial environment better than the natural environment. Historically, this organic chemistry did not develop until the nineteenth century which would seem to leave science from the seventeenth to nineteenth centuries out of the modern epoch. Though Slosson left this ambiguity unresolved, he did not intend to eliminate those two centuries from modernity, of course. Not chemistry itself, but the manipulation of nature has characterized science, and, since Francis Bacon, this has been science's role. This creative power promised to men a true utopia of laborless yet creative and meaningful life. This vision of scientific utopia placed Slosson and the popular science within progressive liberalism's belief that at last the eighteenth-century utopian dream was to come true. [44]

Slosson applauded the "new history" which went beyond traditional political and military events to consider the place of social and scientific developments in man's history. He had the opportunity to write about two of the leading new historians, James Harvey Robinson and H. G. Wells, in the *Independent*. Robinson had defined creative thinking in *The Mind in the Making* (1921) as that process which "makes things look different from what they seemed before and may indeed work for their reconstruction." Robinson illustrated creative thinking, appropriately, with Galileo's discovery of the isochronous period of the pendulum. Discovery alone did not constitute creative thinking, however. The discovery had to enter the social heritage and be

44. Slosson, "Three Periods of Progress," in *Creative Chemistry*, pp. 3, 6–7, 11–13. See also "Back to Nature? Never! Forward to the Machine." See Cushing Strout, "The Twentieth Century Enlightenment," *American Political Science Review* 49 (1955): 321–39. David Noble has pertinent remarks about the content of Enlightenment thought in twentieth-century liberalism, *The Paradox of Progressive Thought* (Minneapolis: University of Minnesota Press, 1958).

developed further by someone else to qualify as creative thinking. Thus, Galileo's discovery had to enter Newton's work. Robinson's concept of creative thinking was simply the same as John Dewey's theory of instrumentalism. It was out of his conviction that ideas caused historical change because of their creativity that Robinson called for a new history, emphasizing the history of ideas. Robinson considered the scientific revolution of the seventeenth century as the crucial event which initiated the modern epoch. It substituted the creative power of the scientific method for the scholasticism and mysticism of the Middle Ages. It was "a historical victory won against extraordinary odds" which has not been repeated.[45]

Slosson admired also H. G. Wells, a personal friend, as a futurist and as a new historian. Wells classified minds according to those which looked to the past and those which looked to the future, calling the latter the *legislative mind*. The legislative mind "thinks constantly and by preference of things to come, and of present things mainly in relation to the results that must arise from them." This type of mind engages in "constructive thinking." Slosson favorably compared Wells's concept of the future-oriented mind to William James's concept of the pragmatic mind. But while Wells called this mind "legislative," Slosson preferred to call it "scientific." Because Wells thought the constructive mind was the scientific mind, the growth of science in the Middle Ages and its culmination in the scientific revolution was of "ultimate importance in human affairs." Science had broken the power of tradition and the past-oriented mind, allowing Western man to create a new civilization. All of this Slosson certainly approved.[46]

45. Slosson, "A Number of Things," *Independent* 84 (November 15, 1915): 288; "Scientific Factors in History," in *Chats on Science* (New York: Century, 1924), pp. 62–66. James Harvey Robinson, *The Mind in the Making* (New York: Harper & Brothers, 1921), pp. 49, 52–53, 146–47, 151–58. Robinson, *The New History: Essays Illustrating the Modern Historical Outlook* (New York: Macmillan, 1912), p. 18. This was the work on which Slosson based his discussion in "A Number of Things" (November 15, 1915).

46. H. G. Wells, "The Discovery of the Future," *Annual Report of the Board of Trustees of the Smithsonian Institution* (Washington, D.C.: Government Printing Office, 1903), p. 375. H. G. Wells, *The Outline of History, Being a Plain History of Life and Mankind*, 2 vols. (New York: Macmillan, 1920), 2: 177. This *seems* to be what Wells was say-

Slosson thought that he had proved that scientific ideas had initiated qualitative, historical change. He found, then, that he was placed on one side of the dilemma he faced when he attempted this proof. Scientific ideas had initiated the modern epoch. The scientific revolution has been the ultimate victory of ideas. But this meant that there had been no qualitative progress since that revolution. The progress achieved since then had been quantitative—the continued rout of the forces of antiscience, such as religion and superstition, and the continued increase of material prosperity and democracy derived from science, for example. But this historical demonstration of the fundamental proposition of the scientists' ideology was an ambiguous victory. For it did not prevent the technologists from arguing that, though scientific ideas had initiated the great cultural movement into the modern era in the seventeenth century, since then progress had been quantitative and dependent on technology. This was, indeed, to be the argument of Lewis Mumford's classic and influential work, *Technics and Civilization* (1934).

Scientific Values in Popular Science

With the fundamental proposition apparently proved, Slosson's popularization could then be directed to the task of relating the values of science to other values of American culture. This attempt to unify science with the rest of culture, and thereby to create a national consensus on science, involved three themes: a definition of the values of science, the relation of scientific values to other cultural values, and the relation of scientific knowledge (as distinguished from scientific technique) to shared social goals.

The values of science were derived from the characteristics of scientific knowledge—unity and law, progress, and the acquisition of scientific knowledge by the "scientific method." The assumption of the unity of scientific knowledge meant, on the simplest level, that true scientific statements never contradict one

---

ing, ibid., 2: 591. Slosson, "H. G. Wells, Social Prophet," *Independent* 76 (November 20, 1913): 349; and "Wells on the World," *Independent* 104 (December 11, 1920): 361.

another. This assumption was synonymous with the assumption of the unity of all knowledge, which, however, had sources in honorable traditions other than science. Hugo Münsterberg expressed one such tradition when he organized the 1904 St. Louis Congress of Arts and Sciences around the theme of the unity of all knowledge. He provided for the general unity of knowledge by a philosophical synthesis which recognized the legitimacy of such diverse intellectual activities as history, ethics, and physics, without reducing any one to another. He had logically distinguished and then synthesized these activities of the mind under their one common pursuit, the progress of mankind. Slosson, who reviewed the proceedings at the congress for the *Independent*, agreed that all aspects of knowledge were unified, but thought a greater vision of their unity was needed than that provided by Münsterberg's philosophical synthesis.[47]

In his popular science, Slosson based the unity of science, not on the unity of all knowledge or on the progress of man, but on the "unity of nature." His interpretation of the unity of nature was laid out in a lecture delivered in 1924, "The Philosophy of General Science." He deplored the specialization of scientific study which prevented a view of the whole of scientific knowledge. The general science course, he thought, could provide a unification of knowledge and a view of the unity of nature. Each phenomenon of nature implied every other phenomenon. The object of scientific research was to discover this unity of nature. Actually, the theory of relativity of Einstein came close to denying that nature has any such intrinsic unity because no action could travel faster than the speed of light, which meant that some phenomena of nature were not touched by others. But Slosson never realized this implication of the theory of relativity. He was convinced that all of nature was pervaded by a "hidden principle . . . call it reason, law, logos, what you will."[48]

47. Hugo Münsterberg, "St. Louis Congress of Arts and Sciences," *Atlantic Monthly* 91 (May 1903): 673–84. Slosson, "A Clearing House of the Sciences," *Independent* 57 (October 6, 1904): 791, 793.
48. Slosson, "The Philosophy of General Science," *School and Society* 20 (December 27, 1924): 799, 801–02, 804.

Exactly what Slosson meant by the "unity of nature" and by the "hidden principle" was not clear. Certainly, by the use of *logos*, he did not mean to interpret modern science in terms of Plato's *Timaeus*, or, by the use of *reason*, to interpret it in terms of Hegelianism. Furthermore, a description of the purpose of science in terms of a search for logos was not in accord with pragmatism. Probably the use of these terms could be traced to vague religious emotions. As a Congregationalist deacon, Slosson was fond of giving a sermon about the "willings" of God as the "immutable laws of nature."[49] This sentiment was, of course, a diluted and genteel form of midnineteenth-century natural theology. The concepts of the unity of nature and the unity of knowledge were, however, also present in Dewey's pragmatism. For Dewey, the distinctions between the categories of knowledge and the distinctions between the categories of nature (for example, organic/inorganic, mind/body) were merely functional. The distinctions were impressed upon the experience of nature to the degree that they aided in the solution of problems.

The unity of science and the unity of nature were secondary values in Slosson's popular science. The primary value was the progressivism of science which was dependent on the scientific method. The scientific method was important, therefore, because it utilized scientific ideas, not because it could be used as a technique in applied science and engineering. Slosson never discussed the scientific method in a systematic way, but mentioned its characteristics incidentally in discussions of scientific discoveries.

There was good reason for Slosson not to analyze the scientific method by itself. It is very difficult, when discussing many different scientific discoveries, to find a single, common, scientific method utilized in all. The method of the biologist working in inheritance, the method of the theoretical physicist, and the method of the behavioral psychologist were quite different, to say nothing of the differences between the methods used by Galileo

49. Slosson, "The Chemistry of the Greatest Miracle in the Bible," *Independent* 55 (June 18, 1903): 1455. This sentiment was repeated in "Faith," in *Sermons of a Chemist*, p. 85.

in the seventeenth century, Darwin in the nineteenth century, and Robert Millikan in this century. It is not to be denied that after discoveries have been made, common characteristics may be found in different methods, but these do not constitute "the scientific method." There are also logical difficulties in merely stating a scientific method. For example, when a method is stated, there is no guarantee that it will result in a discovery or an advance. This obstacle alone would have prevented Slosson from discussing the scientific method because his purpose was to show that the method must result in an advance.

Consequently, Slosson took a psychological approach to scientific method. He emphasized the qualities of mind and personality of the scientist who employed the method. Occasionally, these qualities were confused with the method itself. The scientific method, first, demanded hard work. In one rather clumsy (and unfair) essay, Slosson said that one of the major differences between the sciences and the humanities was that the scientist worked harder, for longer hours, and got his hands dirty. When he described the activity of the scientist, the scientist was always doing something, "always turning over a new leaf [of the book of nature]." By frequent repetition of the theme that the scientific method involved hard work, Slosson, of course, hoped to subvert the popular notion that the scientist was not engaged in a socially productive activity. For the "pioneers" of science, research was "a serious and solemn thing." Science was "toil," "drudgery," and "patient and persistent labor."[50]

What was the objective of this intensive laboring?—the discovery of facts. The discovery of facts also required a special mind. It had to be curious, conformist, and skeptical, all at once. Blended with conformity, the curiosity would yield a "well-considered curiosity." Such well-considered curiosity was, as Slosson put it with unfortunate humor, a cross between motiveless monkeying

50. Slosson, "Science Versus Literature as a Professorial Profession," *Independent* 69 (December 29, 1920): 1440. Slosson, *Keeping Up with Science: Notes on Recent Progress in the Various Sciences for Unscientific Readers* (New York: Blue Ribbon Books, 1924), pp. v–vi. Slosson, "The Chattability of Science," in *Chats on Science*, p. 3. Slosson, "How Scientific Inspiration Comes," in ibid., p. 51.

and aimless aping. Even to see new facts required the scientist to be a skeptic, "for the skeptic is the man who sees, who looks into things, who keeps his eyes open." If the scientist noticed something he suspected to be a new fact, he had to ask, Was it so?[51]

When he had discovered a new fact, the scientist added it to other facts in a "bricklayer" manner, "fact upon fact and cementing them in place with the mortar of logic." The design which the bricklayer-scientist produced was the vision of the rare scientist who saw "the edifice as a whole." This vision needed not to be some logical production, but could come "in a flash quite like the inspiration of the author or artist, and [could come] often when the mind is not consciously working on the problem but is, so to speak, off guard." Slosson did not emphasize the theoretical aspects of scientific research. He wrote that the scientist "holds his theories with a light hand, but keeps a firm grip on his facts." He wished the layman would give less attention to theories and more to facts. To fit facts to other facts in this design, this vision, was to search out the relations between facts. It was, as Slosson emphasized again and again, a process filled with the "hesitations, uncertainties, shyness, [and] professional caution of the true man of science."[52]

There were important reasons for Slosson to emphasize the labor and the facts, rather than the inspiration and the theories, of the scientific mind and method. One reason is that it was fair. The autobiography of most any scientist, and Robert A. Millikan's is typical, contains accounts of much drudgery and repetitive labor—continually refining the accuracy of the electron charge measurements, in Millikan's case—building and rebuilding apparatus, as Ernest Lawrence did the cyclotron. Moreover, American science had a strong nineteenth-century tradition of experimental and observational science, but a weak tradition of

51. Slosson, "Monkeying and Aping," in *Snapshots of Science*, p. 19. Slosson, "The Shut-Eye Skeptic," in ibid., pp. 134–35.

52. Slosson, "How Scientific Inspiration Comes," in *Chats on Science*, p. 51. Slosson, "Religion and Relativity," in *Sermons of a Chemist*, p. 207. Slosson, "Science in Daily Life," in *Snapshots of Science*, p. 94. Slosson, "Science and Pseudo-Science," in *Keeping Up with Science*, p. 3.

theoretical science. One historian has suggested that this weakness in theory was caused by a debilitating adherence to the empiricism of Francis Bacon's philosophy of science before the Civil War.[53] Following the Civil War, the federal government's support of scientific research, as in the geological and mining surveys, pushed scientists into observational and practical science, rather than into theory.[54]

Yet there were exciting visions in science which Slosson chose not to emphasize in his popularization. There was one more reason why Slosson emphasized fact over theory which may explain this choice. This was the prominent place of the literal fact in recent American literature. Although the tradition of philosophical idealism in America had imbued the observed fact with transcendental qualities, this tradition was in decay by the end of the last century. In the absence of idealism, the observed fact began to take a literal and realistic character; that is, the literal fact was not a sign of anything other than itself, did not point to a transcendental form. This literal fact made its appearance in naturalism in the arts and critical realism in philosophy.[55] More important for Slosson's popular science, the literal fact was central to the muckraking journalism of the prewar period.

Robert Bannister's study of Ray Stannard Baker, the muckraking journalist, indicates the special significance of the literal fact. Baker had been an eager student of the sciences while in college. The sciences he learned, however, were observational and Baconian in character. They emphasized careful observation, extensive induction, and the avoidance of generalizations. When Baker began to write muckraking articles (about 1903), the early enthusiasm for science reappeared in detail and handling of facts. He piled up enough facts to point to a conclusion, but without a

53. George H. Daniels, *American Science in the Age of Jackson* (New York: Columbia University Press, 1968), ch. 3.

54. Thomas G. Manning, *Government in Science: The U.S. Geological Survey, 1867–1894* (Lexington: University of Kentucky Press, 1967), ch. 6.

55. Walter Bagehot, in *The English Constitution* (2nd ed., 1872; reprint ed., London, 1905), made the observation that the quality of literality in the American mind was derived from the colonial and frontier experience of overcoming the wilderness (p. 252). I am indebted to Sharon Carroll for this reference.

tight and reasoned argument. Slosson had a great affinity for the muckraking style of journalism and at one time had suggested that Science Service undertake scientific crusading and muckraking. There must have also been in Slosson's mind the memory that the American public had responded warmly to muckraking. Muckraking had helped build a national consensus for reform. Perhaps, he may have wondered, such journalism could help build a national consensus on the importance of science.[56]

Edwin Slosson's popularization of science had demonstrated the importance of science in sociological and cultural change. It had also demonstrated, not always with clear success, that scientific ideas initiated qualitative, historical change. Moreover, the popular science had presented the values of science and illustrated the scientific method in a way that allowed the layman to sympathize with these values. Popular science had carefully made the layman aware that science was closely aligned with American liberalism and the American philosophy of pragmatism.

But there was, in all this, a single great flaw. The popular science presented only one view of the world of science. It did not consider the developments that were revolutionizing theoretical physics. When popularization finally had to face the theories of relativity of Einstein, the sympathies of the layman for science and the carefully knit unity of the two cultures would be shattered once again.

56. Robert C. Bannister, Jr., *Ray Stannard Baker: The Mind and Thought of a Progressive* (New Haven: Yale University Press, 1966), pp. 31, 94. See also note 20 above.

(4) THE EINSTEIN CONTROVERSY, 1919-1924

The experimental facts [of the new physics] are so utterly different from those of our ordinary experience that not only do we apparently have to give up generalizations from past experience . . . , but it is even being questioned whether our ordinary forms of thought are applicable in the new domain; it is often suggested, for example, that the concepts of space and time break down.

P. W. Bridgman, The Logic of Modern Physics

The Special and General Theories of Relativity

It was not obvious that Einstein's theories of relativity should have disrupted the consensus on science which the scientists were building. For several reasons, the theories might well have found a welcome reception, at worst indifference, from the American layman. As a summary and explanation of experimental phenomena, the special principle of relativity, for example, had little direct significance for the consensus. The experimental phenomena were available only to the physicist in the specially equipped laboratory. The theories had no conceivable economic, industrial, or commercial applications which might disrupt the daily life of the public.

The philosophical foundations of the theories of relativity reinforced, rather than conflicted with, the English and American traditions of objective empiricism and realism. The theories did contradict idealistic metaphysics and transcendentalism, but in

America these philosophies had long been in decay and had few adherents. Popularizers who emphasized the psychological aspects of the theories—which might have identified the theories with subjectivism—did not deny that the phenomena of relativity depended on objective, physical situations.

The mathematical incomprehensibility of the theories did not disrupt the consensus on science. Since the progressive period, the lay public had been disposed to accept the incomprehensibility of the expert's techniques. Indeed, the obscurity of the mathematics provoked the curiosity of the layman who wondered what importance such a theory could have. The critics of the mathematical abstruseness of the theories were usually secondary scientists who might have been expected to master the new techniques. Nevertheless, few scientists criticized or rejected the theories on the basis of their mathematical difficulty alone.

Regardless of why the theories might have found a welcome from the lay public, they did not. The theories provoked a controversy of the most profound philosophical issues with a wide public following and participation. From late 1919 through 1924, this controversy raged intensely, seriously weakening the consensus on science and challenging the values of science, which Slosson and the scientists had been popularizing, in two main ways. First, the scientists were required to accept the theories on grounds very different from the grounds on which the lay public could accept them. Second, the layman was unable to understand the theories because the theories made common experience and common sense inadequate for such an understanding. Both of these consequences contradicted what the public had been led to expect from popular science, estranged the public, and thereby undermined the national consensus on science.

Most popularizers and interpreters tended indiscriminately to subsume under the heading "the theory of relativity" all scientific and philosophical matters related to Einstein's theories. Nevertheless, several sets of distinctions can be drawn which clarify what they were writing about. The distinctions of the first set are between the three separate Einstein theories: the special theory of relativity (1905), the theory of gravitation, and the general theory of relativity (1916).[1] The special theory of relativity extended the Galilean principle of relativity, which Galileo limited to mechanical phenomena, to optical and electrical phenomena. Thus, the special principle of relativity stated that no mechanical, optical, or electrical phenomena can be used to detect a difference in natural laws for systems at rest or in uniform motion relative to one another. This principle was one postulate of the special theory. The other postulate was that the velocity of light in a vacuum is a constant. These two postulates in the special theory denied the possibility of detecting experimentally the ether, absolute space, and absolute time. The theory of gravitation described gravity in terms of a curvature of the four-dimensional world (space-time) due to the distribution of matter. Newton had described gravitation as a force acting between masses in a three-dimensional world of space. The general theory of relativity generalized the special theory of relativity to assert that the laws of nature are the same, regardless of the state of motion of the observer (at rest, or in uniform, accelerated, or rotatory motion).

The distinctions of the second set are between the contents of the different theories. These contents are the special principle of relativity and the principle of equivalence (that the force of gravitation cannot be distinguished from uniformly accelerated motion), the philosophical standpoint of the theories of relativity, and the theories of nature inferred from the theories of relativity.[2] The principles of special relativity and equivalence were statements of empirical observations and experimental phenomena. The special principle, for example, was a general summary statement of the negative results of experiments like the famous Michelson-Morley experiment of 1887 to detect the motion of the earth through the ether.[3] Few scientists denied these principles.

---

1. One popularizer who did not confuse these three theories was Morris R. Cohen, "Einstein's Theory of Relativity," *New Republic* 21 (January 21, 1920): 228–31, reprinted in Cohen, *Studies in Philosophy and Science* (New York: Henry Holt, 1949), pp. 215–23.

2. This second set of distinctions was originally proposed by Arthur S. Eddington, *Space, Time and Gravitation: An Outline of the General Relativity Theory* (1920; reprint ed., New York: Harper & Row, 1959), pp. 28–29, 180–82.

3. As a summary statement, the special principle said that "it is impossible by any experiment to detect uniform motion relative to the ether." *Ibid.*, p. 20.

Those denying the special principle did so since they thought they had detected the motion of the earth through the ether.[4]

The philosophical standpoint of the theories of relativity went beyond the principles and rejected traditional physical concepts, substituting new ones if necessary. Many scientists interpreted the special principle to mean that the ether did not exist and therefore rejected the physical concept of the ether. Scientists also interpreted the postulate of the constancy of the speed of light in a vacuum to mean that no physical action, including gravitation, can surpass the velocity of light, though this inference was not first stated by Einstein. Acceptance of this inference meant the rejection of the concept of absolute simultaneity and the concept of absolute (or "real") dimensions. The philosophical standpoint of relativity theory was indebted to the nineteenth-century tradition of critical empiricism of which Ernst Mach was the most important representative. Einstein himself was personally indebted to Mach's views.[5] In the place of the rejected traditional concepts, the philosophical standpoint often substituted others, such as the four-dimensional space-time continuum proposed by H. Minkowski.

Theories of nature inferred from relativity theory concerned ontological questions and were not strictly the province of physics. They raised the question, for example, whether a concept of theoretical physics like the four-dimensional space-time continuum corresponded to a reality outside of physics. They also raised the question of how such a reality was related to the qualitative world of perception. Arthur S. Eddington and Bertrand Russell argued, for example, that nature is not composed of "things" or physical objects, but of events. These events of nature were similar to the qualitative world of our perceptions in which the constituents of our experience are only momentary, forever being created and passing out of existence.[6]

---

4. For example, Dayton C. Miller of Case Institute. See Lloyd S. Swenson, "The Ethereal Aether: A Descriptive History of the Michelson-Morley Aether-Drift Experiments, 1880–1930" (Ph.D. diss., Claremont Graduate School, 1962).

5. See Gerald Holton, "Influences on Einstein's Early Work in Relativity Theory," *American Scholar* 37 (Winter 1967–68): 59–79.

6. Eddington, *Space, Time and Gravitation*, pp. 186–87; Bertrand Russell, *The A B C of Relativity* (New York: Harper & Brothers, 1925), pp. 208–09.

Only in one sense are the distinctions not historically accurate. The special principle of relativity and the special theory of relativity were not designated "special" until 1916, when Einstein announced the general theory. And before 1911 or 1912, the special theory was usually called only the "principle" of relativity.

The Physicists' Reception of the Theories

There were two periods in the American reception of the theories of relativity. From 1905 to 1916, the physicists and mathematicians almost exclusively were concerned with the special theory. From 1919 to 1924, the lay public was engaged in the Einstein controversy, concerned with all the theories. The responses of the physicists to the theories presaged several of the themes that were to occupy the lay controversy.

Extensive historical research into the reception of the special theory in America has not been made, but it is possible to ascertain from published sources that as early as 1908 a group of young scientists at the Massachusetts Institute of Technology was investigating the special theory. Gilbert N. Lewis, associate professor of chemistry, and Richard C. Tolman, instructor in theoretical chemistry, had been developing non-Newtonian mechanics and correlating their deductions with the special theory. Their colleague, Daniel F. Comstock, instructor in theoretical physics, was also exploring the special theory. Lewis and Tolman presented their work to the American Association for the Advancement of Science in December 1908. Tolman quickly became one of the leading relativists. Daniel Comstock delivered several papers on relativity in October and December 1909. At the December 1909 joint meeting of the American Association for the Advancement of Science and the American Physical Society, where Tolman and Comstock presented papers, other papers also concerned with relativity phenomena were delivered.[7] The re-

7. Gilbert N. Lewis, "Non-Newtonian Mechanics," *Physical Review* 27 (December 1908): 525–26; Gilbert N. Lewis and Richard C. Tolman, "Non-Newtonian Mechanics and the Principle of Relativity," read at the meeting of Section B of the American Asso-

sponse of the audience at this meeting was so interested and inquiring that Comstock felt obliged to write a simple expository article on the principle of relativity. In this article, Comstock concluded "that the principle [of relativity] is already in harmony with so many phenomena that the burden of proof lies with those who object to it."[8]

By 1911, as one British physicist noted, the plethora of articles in professional journals signified the intense interest in Einstein's theory and, as well, the extensive investigations on matters related to it. Of course, not all the attention to the special theory was favorable. William Francis Magie, professor of physics at Princeton University and president of the American Physical Society, in an address of December 1911, declared that the special theory was unintelligible. He did not mean that the mathematics of the theory could not be manipulated and brought into some correspondence with experiment. He meant that the theory could not be expressed in terms of concepts like space, time, and force which were derived from immediate experience. These concepts "are the same for all men now, have been the same for all men in the past, and will be the same for all men in the future." The theory had to be intelligible to all men, not just to physicists. Magie's criticisms introduced two of the themes that were to be brought out in the Einstein controversy: that common experience and common sense (concepts like space, time, and

ciation for the Advancement of Science, December 1908, Johns Hopkins University, and published (abstract) in *Physical Review* 28 (February 1909): 150, and (in full) as "The Principle of Relativity and Non-Newtonian Mechanics," in *Proceedings of the American Academy of Arts and Sciences* 44 (July 1909): 709–24; Richard C. Tolman, "The Second Postulate of Relativity," read at the joint meeting of Section B and the American Physical Society, December 1909, Boston, and published (abstract) in *Physical Review* 30 (February 1910): 291, and (in full) in *Physical Review* 31 (July 1910): 26–40. Daniel F. Comstock, "A Neglected Type of Relativity," read at the meeting of the American Physical Society, October 23, 1909, Princeton University, and published (abstract) *Physical Review* 30 (February 1910): 267, and "The Relativity Dilemma," read at the joint meeting of Section B of the American Association for the Advancement of Science and the American Physical Society, December 1909, Boston (apparently unpublished). Will C. Baker, " 'Bound Mass' and the Fitzgerald-Lorentz Contraction," *Science*, n.s. 31 (April 15, 1910): 590; H. A. Wilson, "The Relative Motion of the Earth and the Ether," *Science*, p. 591.

8. Daniel F. Comstock, "The Principle of Relativity," *Science*, n.s. 31 (May 20, 1910): 767, 771.

force) were inadequate to understand the special theory. Magie's criticisms and these themes were shared by other physicists. William J. Humphreys, a physicist at the weather bureau, thought that though the mathematics of the special principle fascinated mathematicians, the theory was "abhorrent to that host of physicists who can no more conceive of time as a function of velocity than they can imagine space to be curved or picture to themselves a fourth dimension."[9]

Very little of this prewar controversy among the physicists gained public attention. Outside the community of scientists, apparently only the *Nation, Monist* (the philosophical journal), *Popular Science Monthly*, and the *Scientific American* carried notices of the controversy. Even in these magazines, however, the authors were scientists or scientifically oriented philosophers. Lewis Trenchard More, a professor of physics at the University of Cincinnati who later wrote a biography of Newton, attacked the special theory in the *Nation* in 1912. More's criticism was that the criterion for truth of the theory was not qualitative experience. He did not think that economy of statement should replace qualitative intelligibility as a requirement of physical theory. "It is better," More wrote, "to keep science in homely contact with our sensations at the expense of unity than to build a universe on a simplified scheme of abstract equations."[10]

The opposition of Magie, Humphreys, and More was not only to the theory of Einstein but to a direction that physics had been taking since the late nineteenth century. Some other physicists were beginning to accept the thesis of the school of critical empiricism of Mach and the school of conventionalism of Henri Poin-

9. E. E. Fournier D'Able, "Principle of Relativity: A Revolution in the Fundamental Concepts of Physics," *Scientific American Supplement* 72 (November 11, 1911): 319 (reprinted from *The English Mechanical World of Science*). Magie, "Primary Concepts of Physics," *Science*, n.s. 35 (February 23, 1912): 293. This was Magie's presidential address to the American Physical Society. Humphreys, "What is the Principle of Relativity?" *Scientific American* 106 (June 8, 1912): 526.

10. L. T. M[ore]., "The Theory of Relativity," *Nation* 94 (April 11, 1912): 371. More was also attacking Planck's quantum theory (which stated that radiant energy was emitted in discontinuous quantities, rather than in continuous waves). *Popular Science's* notice of the new theory was in W. Marshall, "Theory of Relativity and the New Mechanics," *Popular Science Monthly* 84 (May 1914): 434–48.

caré that the critical examination of supposedly fundamental postulates and basic concepts, with the acceptance of economy of thought as a major requirement of physical theory, could bring scientific progress as well as could empirical discoveries. The abandonment of empirical models in physical theory, for example, was one consequence of the critical position. Hypotheses were not intended to relate the theory to qualitative experience like concrete models but to assist calculation, fix ideas, suggest new directions of inquiry, or bring facts together.

The fact that such important critical developments and controversy in physics could go unnoticed in the public press was one more unfortunate indication of the virtual death of popularization of science before the world war. Not only did the periodicals fail to pick up the developments, but the standard reference works in most cases failed to do so. For example, *Webster's International Dictionary* (1901), *Standard Dictionary* (1906), *Webster's New International Dictionary* (1910), and the *Century Dictionary and Cyclopedia* (1913) had no entries which reflected the meanings given by the new physics to *atom, quantum, ether,* and *relativity.* Ruoff's *Standard Dictionary of Facts* (1914) likewise contained no new usages for these words. Nevertheless, the 1910 edition of Webster's did contain a discussion under *atomical* which revealed awareness of J. J. Thomson's theory of atomic and radioactive emissions. The *Americana Supplement* (1911) in the article "Atom" elaborated the most recent developments in atomic theory, making one reference to 1910. Its article "Ether" mentioned Lorentz, Einstein, and Minkowski, and quoted from a 1910 article by Max Planck in which the revolutionary implications of the special theory of relativity were discussed. (The *Scientific American* edited *Americana.*) And the 1916 edition of *The New International Encyclopedia: Courses of Reading and Study* recommended the special theory and the quantum theory in the study of physics, recognizing that these two new theories "severely" tested the older ones.[11] But only a few hints of the

11. *Webster's International Dictionary of the English Language* (Springfield, Mass.: G. & C. Merriam, 1901); *Standard Dictionary of the English Language* (New York: Funk &

new physics were being heard by the lay public before the war. Cultivation of Einstein's theory continued. In 1911, Einstein predicted that rays of light passing close to the sun would be deflected from a straight path, thereby providing the opportunity to verify the proposition derived from the special theory that the propagation of light was influenced by gravitation. Since 1907, Einstein had been working on the general theory. He published his paper on this and on a theory of gravitation in 1916. These two theories were sufficient to account for the advance in the perihelion of the orbit of Mercury, which Newton's law of gravitation could not predict to the observed value, and to predict the displacement of light from large stars toward the red end of the spectrum. The starlight deflection, the derivation of the advance of the perihelion of Mercury, and the "red shift" were to be the three major "tests" of the theories of relativity. In 1917 the Astronomer Royal of England suggested that the solar eclipse of May 29, 1919, would provide an excellent opportunity to verify Einstein's prediction of starlight deflection. Preparations for the expeditions to observe the eclipse were made during the war.[12]

The Popular Reception of the Theories

For the American public of nonphysicists, the Einstein controversy erupted shortly after November 7, 1919, when the Royal Astronomical Society and the Royal Society of London reported that analysis of the photographs taken of the solar eclipse of May 29, 1919, at Sobral, northern Brazil, and the Isle of Principe, west Africa, substantially verified Einstein's prediction of star-

Wagnalls, 1906); *Webster's New International Dictionary of the English Language* (Springfield, Mass.: G. & C. Merriam, 1910); *The Century Dictionary and Cyclopedia* (New York: Century, 1911); Henry W. Ruoff, *The Standard Dictionary of Facts* (Buffalo, N.Y.: Frontier Press, 1914); *The Encyclopedia Americana Supplement* (New York: Scientific American Compiling Dept., 1911); *The New International Encyclopedia: Courses of Reading and Study* (2d ed.; New York: Dodd, Mead and Company, 1916).
    12. Einstein's theoretical papers on relativity are collected in H. A. Lorentz et al., *The Principle of Relativity: A Collection of Original Memoirs on the Special and General Theory of Relativity*, notes by A. Sommerfeld, trans. W. Perrett and G. B. Jeffrey (1923; reprint ed., New York: Dover Publications, n.d.). I have adopted Einstein's wording of the predictions, ibid., pp. 160–64. Eddington, *Space, Time and Gravitation*, p. 114.

light deflection.[13] The controversy thus began in a moment of extraordinary agitation in American history. It shared national attention with the strike of coal workers, the raids of Attorney General A. Mitchell Palmer on suspected alien radicals, the investigations of New York's Lusk Committee on radicalism, the festering disagreement between President Wilson and the Senate critics over ratification of the Versailles treaty, the Russian civil war, and the political struggle over prohibition.

The question arises why the controversy over the theories of relativity did not occur before the war, in response to the controversial reception of the special theory by the physicists. Two answers are obvious. The tradition of popular science within which such a controversy would have had to occur was almost nonexistent preceding the war. Also, the drama of the solar eclipse expeditions was unavailable before the war. But a more adequate answer can be given. A national controversy over relativistic physics would not occur until science occupied a national position in America. The First World War had thrust the physical sciences into this position. After the war the scientists were elaborating an ideology of science and building a national consensus. The controversy occurred because the Einstein theories conflicted with the interpretation of science presented in the popular science by the scientists and undermined the consensus which was to have unified the two cultures. The Einstein controversy did not continue from mere lay curiosity; it was sustained by the profundity of the philosophical and cultural issues it raised.

Within a month of the report of the Royal Astronomical Society and the Royal Society of London on the eclipse, the *New York Times* had published six editorials on the theories of relativity. With noticeable resentment, the editorials expressed the dominant initial reaction of the lay public to the theories and their incomprehensibility. "Quite the most disturbing feature of the situation," the first editorial said, "is the assumption that only

13. The *New York Times* announced the release of the report on the expeditions and the joint meeting of the societies on November 7, 1919, in the story "Eclipse Showed Gravity Variation," November 9, 1919. The most lucid account of the expeditions is in Eddington, *Space, Time and Gravitation*, ch. 7.

men of wonderful learning have the ability, and therefore the right, to see what meaning there is in the fact that light, being subject to a turning from the straight path by a mass of matter like the sun, must itself have of matter at least the quality of weight." Einstein's humorous remark that twelve persons understood the new theories made the *New York Times* only a little happier. If this were the case, no one, not even uncomprehending scientists, could be offended. Failure to understand the theory was not due to a failure of intelligence but to a lack of training. The expert's conclusions had, therefore, to be accepted on authority. One commentator wrote, with thoroughly unintended irony, that "our respect for the human intellect is somewhat increased by the knowledge that there actually are as many as twelve people who do understand the mathematics." No one was sanguine about a situation in which only a score of persons could understand the most important intellectual development of the century. Morris R. Cohen, a respected philosopher at City College, New York, and a personal friend of Einstein, thought that the essence of Western civilization was imperiled: "Free civilization means that everyone's reason is competent to explore the facts of nature for himself; but the recent development of science, involving ever greater mastery of complicated technique, means in effect a return to an artificial barrier between the uninitiated layman and the initiated expert."[14]

The fear that a free and democratic civilization was endangered by a dependence on esoteric experts, whose conclusions had to be accepted on authority, was not original with the outbreak of the Einstein controversy in 1919. This fear had been expressed occasionally during the progressive era when experts came to occupy a strategic position in social thought.[15] But the

14. *New York Times*, November 11, 1919, "Topics of the Times," p. 12; ibid., November 13, 1919, "Sir Isaac Finds a Defender," p. 12; ibid., November 16, 1919, "Light and Logic," III, p. 1; ibid., November 18, 1919, "Nobody Need Be Offended," p. 12; ibid., November 21, 1919, "They Didn't Ignore Refraction," p. 10; ibid., December 7, 1919, "Assaulting the Absolute," III, p. 1. John Q. Stewart, "The Nature of Things: Einstein's Theory of Relativity—A Brief Statement of What It Is and What It Is Not," *Scientific American* 122 (January 3, 1920): 10. Cohen, "Roads to Einstein," *New Republic* 27 (July 6, 1921): 172–74; reprinted in Cohen, *Studies in Philosophy and Science*, pp. 239–40.

15. See, for example, J. Lee, "Democracy and the Expert," *Atlantic Monthly* 102 (November 1908): 611–20.

progressives' anxiety over the expert was different from that of
the layman in the Einstein controversy. The progressive did not
doubt the ability of every man to understand science and reality.
Every man, however, did not have the time to understand them.
The expert was accepted in terms of social efficiency. Certain
social functions were entrusted to persons whose scientific training was supposed to guarantee democratic justice. In the Einstein
controversy, however, it was precisely the layman's ability to
understand science and reality, regardless whether he had the
time, that was doubted. The expert's conclusions were accepted
on authority, not as a matter of trust and convenience, but out of
necessity.

This new situation was revealed in the manner in which the
popular press was forced to report the anti-Einstein campaign of
Charles Lane Poor, professor of mathematical astronomy at
Columbia University, and Thomas Jefferson Jackson See, a naval
astronomer and mathematician based at Mare's Island naval
yard. Poor did not think that accounting for the advance in the
perihelion of the orbit of Mercury and the deflection of starlight
from a straight path as it passed the sun were sufficient or necessary evidence for Einstein's general theory. Poor argued that with
certain ad hoc assumptions, Newton's law of gravitation would
explain the same phenomena. In this case, Einstein's general
theory would not be logically necessary, that is, the only theory
capable of accounting for the facts. Poor proposed to explain the
deflection of starlight by refraction of the light through the co-
rona of the sun. This was not, however, a scientifically responsible
hypothesis. Robert A. Millikan had suggested the possibility of
refraction earlier, in November 1919. The *New York Times* had
accepted the refraction hypothesis with a sigh of relief: "People
. . . who have felt a bit resentful at being told that they couldn't
possibly understand the new theory [of general relativity], even if
it were explained to them ever so kindly and carefully, will feel a
sort of satisfaction on noting that the soundness of the Einstein
deduction has been questioned by R. A. Millikan." But both
Millikan and the *New York Times* had been quickly humbled.
The gaseous outer atmosphere of the sun was nowhere dense
enough to refract starlight to the observed value of the bending, a

fact Millikan should have checked. The *Times*'s endorsement of the refraction hypothesis was quickly withdrawn. Refraction was finally and decisively refuted in the year previous to Poor's rehabilitation of it by Arthur S. Eddington.[16]

Poor's ad hoc hypothesis to account for the advance in the perihelion of Mercury's orbit was more responsible. He proposed that the advance of the perihelion could be explained by deducing the gravitational force necessary to produce it from the circumsolar lens (the planetesimal bodies dispersed in the solar system). Poor's circumsolar lens hypothesis was given no credence in the scientific community. As J. Malcolm Bird, the Einsteinian editor of *Scientific American*, pointed out, most physicists preferred Einstein's generalized theory to Poor's ad hoc hypothesis. Nevertheless, the popular press thought it had to report Poor's theory. Physics was in upheaval. Having no means to discriminate between responsible and irresponsible science, the press dutifully reported both.[17]

This receptivity opened the popular press to exploitation by the unscientific and unscrupulous. This was particularly the case with T. J. J. See. See's attacks on Einstein erupted into minor controversies in the columns of the *New York Times* in 1923 and 1924. See accused Einstein of plagiarism.[18] He pointed out that the eighteenth-century English scientist, Henry Cavendish, had predicted, as a deduction from the Newtonian law of gravitation and

16. The *New York Times* first reported Poor's views in "Poor Says Einstein Fails in Evidence," February 8, 1921, p. 17. The most thorough presentation of Poor's views, including a defense of the refraction hypothesis, is in his *Gravity Versus Relativity* (New York: G. P. Putnam's Sons, Knickerbocker Press, 1922). *New York Times*, November 12, 1919, "Nogochi Tells [of] Discovery," p. 16; ibid., November 13, 1919, "Sir Isaac Finds a Defender," p. 12; ibid., November 21, 1919, "They Didn't Ignore Refraction," p. 10. Eddington, *Space, Time and Gravitation*, pp. 120–21.

17. Charles Lane Poor, "Planetary Motions and the Einstein Theories: A Possible Alternative to the Relativity Doctrines That Would Save the Newtonian Law," *Scientific American Monthly* 3 (June 1921): 484–86. Poor's views were explained also in his articles, "Eclipse Casts Doubt on Relativity," *New York Times*, January 6, 1924, VIII, p. 4; and "Is Einstein Wrong?—A Debate: I. The Errors of Einstein," *Forum* 71 (June 1924): 705–15. The Einstein editor, "An Alternative to Einstein: How Dr. Poor Would Save Newton's Law and the Classical Time and Space Concept," *Scientific American* 124 (June 11, 1921): 468.

18. The editors of the *Scientific American* had made a general reply to the charge of plagiarism in an editorial, 124 (May 14, 1921): 382.

corpuscular theory of light, that starlight passing close to the sun would be attracted from a straight path. In 1801 the German physicist, J. von Soldner, had derived a quantitative prediction of this deflection similar to Einstein's. See asserted that Einstein merely had patched up von Soldner's formulas and claimed them as his own. Einstein could not be considered a scientist of real note.[19] When the *New York Times* printed the story of See's opinions in April 1923, a letter controversy began which the paper carried for weeks. The controversy faded in August without a resolution of the issues involved.[20] What is instructive about this controversy is that the *New York Times* printed these letters from persons without high standing in the scientific community whose views were presented in an accusatory manner without extended reasoning or evidence.

The See controversy was reopened in the fall of 1924. On this occasion, the *New York Times* took action to squash the controversy before the editorial page filled with opinionated letters. In an address to the California Academy of Sciences on October 13, 1924, See claimed that he had made a "correct" deduction from the Newtonian law of gravitation which yielded a prediction of the starlight deflection more accurate than Einstein's prediction. Furthermore, See imputed that Einstein had made an error in his calculations. According to See, the major astronomical phenomena supposedly verifying Einstein's general theory were actually explained better by Newton's law. Apparently in an attempt to

19. "Prof. See Attacks German Scientist," *New York Times*, April 13, 1923, p. 5. Although never a well-known scientist, T. J. J. See had been eulogized by a biographer as another Herschel who had made unparalleled discoveries and founded two new sciences. W. L. Webb, *Brief Biography and Popular Account of the Unparalleled Discoveries of T. J. J. See, A.M., Lt.M., Sc.M. (Missou.); A.M., Ph.D. (Berol.); Famous Astronomer, Natural Philosopher, and Founder of the New Sciences of Cosmogony and Geogony* (Lynn, Mass.: Tho. P. Nichols & Son, 1913).

20. In chronological order, this little flurry of letters went: Dr. Harris A. Houghton, letter to editor, *New York Times*, April 21, 1923, p. 10 (anti-Einstein); William Marias Malisoff, ibid., May 9, 1923, p. 18 (pro-Einstein); Frederick Drew Bond, ibid., May 13, 1923, VIII, p. 8 (pro-Einstein); Arthur Pestal, ibid. (pro-Einstein); F. D. B., ibid. (thinks See is amusing); Arvid Reuterdahl, ibid., June 3, 1923, VIII, p. 8 (reply to Bond's letter); Frederick D. Bond, ibid., July 15, 1923, VII, p. 8 (reply to Reuterdahl's letter); Reuterdahl, ibid., August 12, 1923, VIII, p. 8 (reply to Bond's of July 15). Reuterdahl was the president of Ramsey Institute of Technology, St. Paul, Minn.

close its columns to a letter controversy, the *Times* asked leading scientists who had made contributions to relativistic physics to comment on See's address. Luther Eisenhart of Princeton University decisively refuted See's Newtonian explanation of starlight deflection. Von Soldner's prediction called for an acceleration of light through the sun's gravitational field; Einstein's prediction was of a deceleration. Einstein could hardly have stolen his prediction from von Soldner. Before he had formulated his own theory of gravitation, Einstein in 1911 predicted a starlight deflection of 0.84 seconds of arc on the basis of Newton's law. Einstein knew this was inaccurate, however, formulated his own theory of gravitation, and later predicted (accurately) a deflection of 1.68 seconds of arc on the basis of this theory. The fact that the observed value of deflection corresponded to the second prediction did not mean, as See claimed, that the first prediction was a miscalculation, but that the first prediction was based on an inadequate theory. The *Times* went also to Arthur S. Eddington, Britain's leading advocate of Einstein's theories. Eddington pronounced See's claim that Einstein had made an arithmetical error as " 'all bosh and nothing to it.' " Talking to a representative of the *Times*, Sir Frank Dyson, Astronomer Royal and secretary to the Royal Society of London, also denied See's allegations against Einstein. After seeing a *Times*'s copy of See's accusation of error, Einstein himself could only remark quizzically, " 'Too bad.' "[21]

The *New York Times* did not have an ideally scientific method of handling See's accusations. The *Times* countered a minor authority with major authorities but without the usual rules of evidence. Persons with an intellectual stake in the theory could hardly have been expected to give See's criticisms—preposterous as they were—a judicious hearing. Support for See's views was not presented along with the denials of his views. Only in Eisen-

21. "Prof. See Declares Einstein in Error," *New York Times*, October 14, 1924, p. 14. The story was covered also, with quotations, in "Is Einstein's Arithmetic Off?" *Literary Digest* 83 (November 8, 1924): 20–21. "Denies See Proved Einstein Wrong," *New York Times*, October 16, 1924, p. 12. "Denies Error in Relativity," ibid. During the academic year 1924–25, Eddington was teaching at the University of California, Berkeley. "Defends Einstein Against Capt. See," *New York Times*, October 16, 1924, p. 12. "Einstein Ignores Capt. See," ibid., October 18, 1924, p. 17.

hart's letter to the *Times* did extensive reasoning back a conclusion. Yet this method of pulling authority down on authority illustrated the peculiar position in which the lay public had been placed by the theories of relativity. The layman was no more able to follow the attack of the anti-Einstein campaign than it was able to follow the defense of Einstein's advocates. Rules of evidence and extended reasoning behind conclusions were irrelevant for the layman since he could understand neither. The layman, then, could only keep score on the number of scientists, and persons who claimed to understand the theories, who were for or opposed to the theories. This was part of the reason why the *New York Times* for so long kept its columns open to "debate" over the theories.

It is not fair to say that the Einstein critics like Poor and See sought to exploit the receptivity of the popular press simply out of maliciousness or a desire for personal publicity. There was enough evidence of Poor's scientific sincerity and concern for public opinion to warrant his place in the popular pages, though this cannot be said of T. J. J. See. More important, however, these men sought out the popular press because they were denied a hearing in the professional, scientific journals. (They were not denied a hearing altogether because they occasionally gave lectures at scientific societies.) See's major refutation of Einstein's theories—accepting his own word on this—remained an unpublished typescript which he sent in 1925 to the Berlin Academy of Science. Poor complained that the scientists in control of the National Academy of Sciences and the scientific journals refused to investigate the alleged errors of Einstein and to print any criticisms of the theory.[22]

For all of this, there was a more profound reason why the newspapers and popular press were receptive to all classes of

---

22. See's treatment was entitled "Research in Non-Euclidean Geometry and the Theory of Relativity: A Systematic Study of Twenty Fallacies in the Geometry of Riemann, Including the So-Called Curvature of Space and Radius of World Curvature, and of Eighty Errors in the Physical Theories of Einstein and Eddington, Showing the Complete Collapse of the Theory of Relativity." See his letter to the editor in the article "See Says Einstein Has Changed Front," *New York Times*, February 24, 1929, II, p. 4. Charles L. Poor, letter to editor, ibid., September 5, 1926, VII, p. 14.

opinion on the controversy and why they took the approach of counting authorities pro and con. The lay public wanted to know whether the theories were true. The layman was waiting for the scientists to agree on the truth of the theories. Until agreement was reached, the press kept watch on the division of scientific opinion. This was the only approach the layman could take to the theories and the controversies, for he had no use for the theories unless they were true. If they were true, he might have to alter his beliefs about science or the universe. But this approach of the layman to the theories was irrelevant to the scientist. It indicated how little the layman understood about the revolution going on in physics. The scientist was not concerned about the truth of the theories; yet he had many uses for them. For example, the theories provided simple, mathematical explanations for a number of crucial experiments. Applications could be made to fields not originally covered by the theories, like Arthur Sommerfeld's application of them to the very high orbital motions of electrons. The theories advanced the progress of science by suggesting new areas of research, thereby holding out the possibility of new discoveries. To all of this, the "truth" of the theories was irrelevant. They could have been "false" and achieved the same. If the scientists had not accepted the theories of relativity because they believed them to be "false" and had accepted instead Poor's and See's ad hoc hypotheses because they believed these to be "true," they would have deprived themselves of more economical physical statements, new experiments, and new areas of research. For the layman, of course, these uses of the theories were of no concern. The layman did not engage in scientific activity and had no reason to accept a theory merely because it suggested additional activity.

This major distinction between the layman's reception and the scientist's reception of the theories of relativity appeared clearly in the responses of scientists to a debate on relativity between Charles L. Poor and Archibald Henderson, a professor of mathematics at the University of North Carolina who had made contributions to relativity. Many scientists responding to the debate considered the question whether Einstein's theories were true or false to be the wrong one. They preferred to consider the theory as "successful" or "sound" rather than "correct," "right," or "true." Joseph Seidlin of Alfred College expressed this sentiment well: " 'By a successful theory is meant one which explains certain facts, which inspires and encourages thinking men to carry on in their respective fields with added vigor, which recreates and enlarges our conception of the universe, which gives birth to numerous tributaries, and which arouses the otherwise sluggish public.' " In a long quotation, Phillip Franklin explained that the two requirements of a physical theory were that it accounted for the facts and did so simply. Truth, meaning correspondence to eternal forms of common experience of nature, was not a requirement. " 'Eternal verity has no place in our philosophy [that is, science]. From this point of view Einstein's theory is correct because its postulates are sound.' " In his reply to Poor's attack on Einstein's general theory, Henderson enforced this point with the statement that "the validity of relativity is not *proved* by the three experiments [accounting for the observed value of the advance of the perihelion of the orbit of Mercury, the deflection of starlight from a straight path as it passes the sun, and the red shift]: they serve to check the soundness of the fundamental postulates."[23]

Doubtless this must have confused the lay mind. Scientists were not interested in whether the Einstein theories were "true" but whether they were successful. The much-publicized solar eclipse observations and the other experiments were not intended to prove the truth of the theories. Doubtless this seemed like an evasion of the criticisms of Poor, See, and earlier critics like Magie, Humphreys, and More, who had claimed that the theories did not fit the facts any better or, as the latter three had charged, relate to our common experience and common sense as well as

23. Charles L. Poor, "Is Einstein Wrong?"; Archibald Henderson, "Is Einstein Wrong?—A Debate: II. The Triumphs of Relativity," *Forum* 72 (July 1924): 13–21; Poor, "Triumphs of Relativity: Reply," ibid. (August 1924): 273–74; "Is Einstein Wrong? A Symposium: Summarizing or Quoting Opinions of Many Scientists on a Subject Which has been Debated by Charles Lane Poor and Archibald Henderson in the June and July Numbers of *The Forum*," ibid., pp. 277–81. Archibald Henderson was also a historian and a Shakespearean scholar.

Newton's science. The Einsteinian proponents replied that Einstein's theories were more successful. The consensus of the scientists that the "truth" of the theories was not an issue in their acceptance meant that their possible falsity was not the reason for their incomprehensibility to the lay mind.

The layman and the physicist had to judge the theories of relativity by different standards. For the layman and for the scientists who did not accept them, the theories would be accepted if they were true. And the truth of the theories should mean their correspondence to the common experience and the common intuitions of men. For the physicist, however, the theories were judged in terms of their success. The attempt to judge the theories as true in the eternal categories of common experience and common sense would only make the theories incomprehensible. Not being a scientist, the layman had no choice but to judge the theories in terms of common experience and concepts.

Notwithstanding the incomprehensibility, impracticality, and irrelevance of the theories to truth, the layman was assured that they had produced a scientific revolution he was obligated to understand. In 1917, even before the lay controversy began, a French physicist warned in *Scientific American* and *Current Opinion* that science was in revolution. Edwin Slosson often described the theories of relativity as revolutionary—as revolutionary as the achievements of Copernicus and Newton. *Scientific American* considered the theories to be the greatest scientific advance since Newton. Morris Cohen thought the essence of Western civilization could be preserved only if every man were educated to some understanding of the new theories.[24]

The effort to popularize the theories, motivated by the sense of cultural peril of Morris Cohen as well as a recognition of their scientific importance, encountered the ironic difficulty that the theories made the common experience and common sense of men

24. Alphonse Berget, "Principle of Relativity," *Scientific American Supplement* 83 (June 30, 1917): 411; "A New and Revolutionary Doctrine of Time and Space," *Current Opinion* 63 (September 1917): 178. Edwin E. Slosson, "Einstein's Reception," *Independent* 105 (April 16, 1921): 401; "New Wonders," ibid., 108 (May 13, 1922): 444. Stewart, "The Nature of Things," p. 10. Cohen, "Roads to Einstein," pp. 239–40.

inadequate for their comprehension. Thus, popularization sub-
verted the purpose for which it was intended. Popularization was
to have elicited the layman's sympathy for the values of science
and unified the two cultures; instead, it showed that the theories
of relativity made this impossible.

The Inadequacy of Common Experience

In one of the better articles on the theories of relativity, "A
New Conception of the Universe," Alfred J. Lotka explained why
the theories of relativity were necessary. The aim of science, he
began, was "to build up a conception of the world which shall
correspond more and more closely with our experience." Relativity theory was necessary because the "scope of our experience"
had been recently enlarged. What were the new experiences that
demanded relativity theory? Lotka went on, the new experiences
were similar to the common experience that "the aspect of
things" changed when the position of the observer changed.
However, the new experiences were not exactly like this. "The
changes of aspect" with which relativity theory dealt were "of a
particularly baffling character." They were not ordinary experiences.[25]

Most popularizers of relativity were led into this self-contradic-
tion, that though physics ultimately has the purpose of explaining
the experience of men, the relativity theory explained experiences
men did not have. The popularizer could arrive at this self-con-
tradiction by explaining how relativity theory denied absolute
simultaneity, as did Edwin Slosson in "That Elusive Fourth
Dimension" and George D. Birkhoff of Harvard in "The Origin,
Nature and Influence of Relativity," or by explaining how relativity theory used the concept of four dimensions, as did Lotka.

An experience common to all men is the experience of absolute
simultaneity. With few exceptions, sensory perceptions of events
carry the quality of time by which all events occur at the same
instant, that is, the instant of their perception. The author's per-

25. Alfred J. Lotka, "A New Conception of the Universe," *Harper's Magazine* 140
(March 1920): 477.

ceptions of the events in his study, for example, carry this quality of absolute simultaneity. The wind blowing the curtains away from the opened window, the odor of pipe smoke, the words on the paper in his typewriter, as well as the succession of his thoughts about them, occur at the same time. This quality of "nowness" which he perceives in his immediate neighborhood, he perceives also in the whole universe. The radio broadcast from New York City, the blueness of the sky, and the sunlight from the sun happen "now." Only on a few occasions do the senses deny this simultaneity. For example, when lightning and thunder are not perceived simultaneously, the mind is forced to account for the elapsed time between them. But even in this exception, the effort of the mind is to restore simultaneity.

As Birkhoff had to explain, however, according to the special theory of relativity, the commonly experienced quality of absolute simultaneity was not a characteristic of the whole universe. There was no single instant, no "now," shared by all natural events. This is, of course, a direct contradiction of experience.[26] Any attempt to deny the contradiction led to distortions of the special theory. Slosson, for example, attempted to evade the contradiction by asserting that the relativity of time was in accord with common experience. Certainly it was in accord with experience that one's own sense of elapsed time changed with situations. When waiting for a date with a girl friend, the minutes seem to drag, as Slosson said, but when with her, the time seems to fly.[27] But the special theory's denial of absolute simultaneity and assertion of the relativity of time had nothing to do with this psychological characteristic of consciousness. The special theory asserted that our perception of someone else's time, not our own time, was relative, that is, that to our perceptions, his moments of elapsed time were not synchronous (congruent) with our own. This relativity was due to the complexities of measurement and not to the character of consciousness.

26. George D. Birkhoff, "The Origin, Nature and Influence of Relativity: II. The Nature of Space and Time," *Scientific Monthly* 18 (April 1924): 408. This series of articles was also published as a book with the same title in 1924.

27. Slosson, "That Elusive Fourth Dimension," *Independent* 100 (December 27, 1919): 298.

This contradiction of common experience was also achieved by the concept of the four-dimensional space-time that was utilized in the general theory. The four-dimensional space-time was a physical concept proposed by the German mathematician H. Minkowski in an address before the Eightieth Assembly of German Natural Scientists and Physicians in September 1908.[28] Minkowski suggested that physical laws could be expressed in a perfect, absolute way as the relations between what he called "world-lines." World-lines were the paths or distances described by point-particles (substances) moving from one position x, y, z, t, to another position, x', y', z', t', where x, y, and z were spatial coordinates and t was the time coordinate. In Minkowski's scheme, the t variable (time) had no distinction from the other three variables. The value of time was relative, just as were the values of the spatial variables. Thus, time was being treated as if it were a spatial dimension also. Time was an imaginary dimension, however. This imaginary character of the time dimension was a direct consequence of the special theory (specifically, of the Lorentz transformation). Time would not be perceived as another spatial dimension, if men were to have another spatial sense, in the way that "real" three-dimensional space is perceived; this is an epistemological matter, however, and not properly the domain of physics.

The world-lines of a particle would be a curve, and the geometry of such curves would be a four-dimensional geometry. It was possible for Minkowski to describe such relativistic phenomena as contraction of dimensions of objects in terms of the mathematics of his theory. Contraction thus appeared as a natural consequence of a world of four dimensions and lost its physical peculiarity, that is, was not due to the special pressure of an ether or to an electromagnetic force.

This conception of the space-time continuum was derived by Minkowski in part from the denial of absolute time in the special

28. The phrase *physical concept* in this sentence means "a concept used in physics." Such a concept can be mathematical and have no reference to the constituents of men's experiences of the physical world. The address, "Space and Time," is reprinted with notes in Lorentz et al., *The Principle of Relativity*, pp. 75-96.

theory of relativity. Einstein, in turn, was to borrow Minkowski's concepts of space-time and a four-dimensional world to formulate his general theory of relativity. In the latter, for example, the concept of force was rejected. Thus the deflection of starlight from a straight path as it passed the sun appeared not as a result of the attractive force of the sun, but as the natural, curved path of a point-particle moving in a four-dimensional world.[29]

The concept of a four-dimensional space-time was enormously difficult to explain in any terms other than those in which it was formulated. Mathematics, as the popularizers were fond of saying, was as difficult to translate as music. There were, however, several popularizations of the four-dimensional space written before the Einstein controversy to which relativity popularizers turned. These were Edwin A. Abbott's *Flatland: A Romance of Many Dimensions* (1885) and the *Scientific American* prize contest collection, *The Fourth Dimension Simply Explained* (1910).[30]

29. I have presented what was the dominant interpretation of time as a fourth dimension, that is, that time was a dimension similar to space. This was the interpretation of Einstein, Minkowski, Eddington, and most popularizers. Nevertheless, cogent arguments were presented by Alfred North Whitehead and Henri Bergson that this was not the correct interpretation of special relativity. Whitehead and Bergson argued that time as a fourth dimension meant the temporalization of space, that is, that the spatial dimensions took the character of time or duration. This was not a quibble over emphasis, but a serious question about the character of physical reality. The Minkowski-Eddington interpretation leads to a four-dimensional physical reality in which past, present, and future events are already drawn in the world-lines. The Whitehead-Bergson interpretation leads to a physical reality of which temporality is the main characteristic, that is, a physical reality in which the future is not already drawn but is continually novel. Eddington, for example, discussed the possibility of a man equipped with different sensory faculties which enabled him to know the future in the same way we know the past (Eddington, *Space, Time and Gravitation*, pp. 51–52). For Whitehead and Bergson, this would be impossible.

A discussion of these interpretations, which I cannot properly undertake here, is provided by Milič Čapek, *The Philosophical Impact of Contemporary Physics* (Princeton: Van Nostrand, 1961). Čapek takes for his own the interpretation of Whitehead and Bergson, suitably modified.

It might be remarked that Whitehead's philosophy of nature is worth historical investigation. His philosophy was a powerful synthesis of the theory of evolution and the theory of relativity. What reception this synthesis received in America in the 1920s, and why the synthesis was not more influential, would reveal much about the intellectual milieu of the interwar period. The philosophy of nature of John Dewey (as set out in *Experience and Nature* especially) bears strong resemblance to Whitehead's and a comparative study of the two should also be valuable.

30. [Edwin Abbott Abbott], *Flatland: A Romance of Many Dimensions* (1st ed., date not established, 2nd rev. ed., 1885; reprint ed., New York: Dover Publications, 1964). The

Abbott's *Flatland* is a completely charming fiction intended to impart an understanding of the multidimensional character of space. The story concerns the experiences of an inhabitant of a world of two dimensions (breadth and depth, that is, plane) named A. Square, who is the narrator. A. Square describes the world of two dimensions, Flatland, its houses, people, and social order. The inhabitants are all plane figures, like straight lines (who are women), triangles (soldiers), squares (members of the upper class), and circles (priests). He describes the knowledge they have of their two-dimensional world and how that knowledge is obtained. For example, the inhabitants are able to distinguish between one another even though they can only perceive directly straight lines and points. Why this is so can be illustrated. When viewed on a plane, paper cutouts of geometrical figures like triangles and circles look only like straight lines of different lengths. To see the shape of the cutouts one would have to be above or below them, that is, out of the two-dimensional plane. But it is impossible for the Flatlanders to move out of the plane. Consequently, according to A. Square, the inhabitants of Flatland distinguish one another mainly by touch, inferring the shapes of the figures. Thus, the Flatlanders perceive only one dimension of space (breadth), and have inferred knowledge of the other (depth) and the figures which breadth and depth can make.

The Flatlanders have no direct perception of the third dimension (height) and no inferred knowledge of the existence of a three-dimensional world (one in which solid figures can exist). Thus, if someone in the three-dimensional world were to poke a cylindrical rod into the Flatlanders' two-dimensional world, the Flatlanders could not perceive that it was a rod. They would infer that it was a circle or other closed-curve figure at the place where their two-dimensional plane cuts the rod. The Flatlanders are able, however, by the method of mathematical, analogical reason-

---

second revised edition carried the delightful pseudonym, "A. Square." Henry P. Manning, ed., *The Fourth Dimension Simply Explained: A Collection of Essays Selected from Those Submitted in the Scientific American's Prize Competition* (1910; reprint ed., New York: Dover Publications, 1960).

ing to understand that mathematically there could be more than a two-dimensional space. The analogy would proceed this way: The Flatlanders understand that a point can generate a line by moving through a plane. The line can generate a two-dimensional figure by moving parallel to itself in the plane. Now, the Flatlander can ask, could not some figure (for which the Flatlander has no name, which he could not perceive directly, and of the existence of which he could not have inferred knowledge) be generated by a two-dimensional figure moving through space not already occupied by itself, that is, out of the plane?

The rest of Abbott's story tells how a being from the world of three-dimensions (Spaceland) visits A. Square in Flatland and, by various miracles, transports him to the world of three dimensions. The inhabitant of the three-dimensional world has the ability to perceive directly two dimensions (breadth and height) and to infer knowledge of the existence of the third (depth) and thus knowledge of solids. By more miracles, the inhabitant of the three-dimensional world gives the Flatlander hero the ability to infer the existence of solids, thus proving to the Flatlander that a world of three dimensions exists.

Flatlander's experiences in his native two-dimensional world and in the miraculous three-dimensional world give him the impulse to ask whether a four-dimensional world exists. Flatlander reasons by analogy, thus: There is a world of two dimensions, whose inhabitants perceive directly one dimension and infer the other; there is a world of three dimensions, whose inhabitants perceive directly two dimensions and infer the other. Could there not be a world of four dimensions, whose inhabitants perceive directly three dimensions and infer the other? Just as the spatial dimensions of the two-dimensional world are generated by a line moving parallel to itself in a plane and the spatial dimensions of the three-dimensional world are generated by the points of a plane figure moving through space not previously occupied, so the spatial dimensions of the four-dimensional world would be generated by the points of a solid moving through space they have not previously occupied. Such a moving solid would generate a hypersolid. Just as it is possible for a being in the three-dimensional world to see the inside of the two-dimensional figures, of which Flatlanders can see only the outside, so it would be possible for an inhabitant of the four-dimensional world to see inside the three-dimensional, solid figures of which the Spacelander can see only the outside.

Abbott's story of Flatlander's adventures in Spaceland became a model for the popularization of the concept of four-dimensional space-time. The popularizers did not expect that the lay public could be led to an understanding of the four-dimensional world by an abstract discussion alone and so were forced to discuss the four-dimensional world in terms of actual or imaginary experiences.[31] For two reasons, however, the Flatland model was inadequate for the tasks of explanation which the theories of relativity demanded. First, the method of analogy used by Abbott to lead the reader to a conceptual understanding of the geometrical properties of four-dimensional space could not also be used to lead the reader to an experiential understanding of four-dimensional space. Second, neither a conceptual understanding of the geometrical properties of four-dimensional space nor an experiential understanding of four-dimensional space (if this were possible) would lead the reader to an understanding of four-dimensional space-time.

The first reason can be established in the following obvious manner. The prizewinning essay in the *Scientific American* competition on popular expositions of the fourth dimension of space, "An Elucidation of the Fourth Dimension," by Lt. Col. Graham Denby Fitch, U.S.A., began with the statement that "it is impossible to form a mental picture of the fourth dimension." Another writer declared, "The fourth dimension has no real existence in the sense in which the external world that we know by means of our senses has real existence."[32] Human beings do not

31. Eddington was the leading proponent of the view that a real four-dimensional world exists. See the final chapter of *Space, Time and Gravitation*. His book undoubtedly influenced the popularizers to try to describe the experiences one would have in such a world.

32. Lt. Col. Graham Denby Fitch, "An Elucidation of the Fourth Dimension," in *The Fourth Dimension*, p. 43. Edward H. Cutter, "Fourth Dimension Absurdities," in ibid., pp. 60–61.

have the sensory faculties to experience a fourth dimension. Therefore, it is impossible to predict or even to imagine what the fourth dimension would be like if humans had the sensory faculties to experience it. This implication follows because it is not possible on the basis of experience to predict or to imagine what a qualitatively different experience would be. Thus, it is impossible from the experience of vision to predict or to imagine what the experience of hearing would be like, if one had never had the experience of hearing. In Abbott's *Flatland*, for example, this analogy of experience is made: that just as an inhabitant of the three-dimensional world can see inside the plane figure of a two-dimensional world, so the inhabitant of the four-dimensional world could see inside the solids of a three-dimensional world. There is no true analogy here at all. These two visions are not in the same category of experience. First, no one knows what it would be like to "see" in a four-dimensional world. No one knows "what" would be seen. Second, a being in a four-dimensional world would not see inside solids because there would be no solids in his world, only what are called hypersolids. And who knows what a hypersolid looks like? Analogies cannot be drawn from items of which men have experiences to items which they cannot possibly experience.

These limitations of analogy in popular science were further revealed in illustrations of the fourth dimension. One mathematical property of a four-dimensional space is that rotation can occur around an axis plane, whereas in three-dimensional space, rotation can only occur around an axis line. This property of rotation about an axis plane provides a mathematical technique for explaining and manipulating such physical phenomena as the change of right-handed polarization of light into left-handed polarization of light.[33] The attempt of the popularizers to provide concrete illustrations for this abstract conception of rotation around an axis plane led to hopeless confusion and further revealed the inadequacies of common experience. Thus, Slosson

33. Fitch, "An Elucidation of the Fourth Dimension," in *The Fourth Dimension*, pp. 47–50.

told how a right-handed glove turned over through four-dimensional space becomes a left-handed glove. In four-dimensional space, a string held at both ends could still be knotted or unknotted.[34] No meaning can be made of these illustrations. They contradict the admitted fact that the fourth dimension cannot be experienced. A glove cannot be turned into its mirror image because the fourth dimension is inaccessible to gloves. If the fourth dimension were accessible, our three-dimensional glove would no more look like it does to us in a three-dimensional world than a square looks like a cube. To be meaningful, any proposition or illustration must be testable, and no one can test an illustration or proposition about what his experiences, with gloves, strings, or whatever, would be in a four-dimensional world.

The second reason for the inadequacy of the Flatland model was that popularization of four-dimensional *space* could not give any understanding of four-dimensional *space-time*. The attempt to understand the latter through the former involved a confusion in the popularizers' minds. Although time was relative, like space, in the special theory, time-the-fourth-dimension was not— as the popularizations of it would lead the layman to believe— a fourth spatial dimension. Indeed, what the special theory did was to make time, which was not previously considered a dimension, into a dimension in its own right, so to speak. In the special theory, the time dimension is not real as the three spatial dimensions men do experience are real. The fourth, time dimension, is not experienced. Furthermore, if the fourth, time dimension, were real in some sense, men could not have actual or imaginary experiences of it for the same reason men could not have such experiences of a fourth spatial dimension.

The method of mathematical analogy used to construct a geometry of four dimensions had no parallel method of analogy to construct the experiences a person would have in a four-dimensional world. The method of analogy could not, then, lead the layman from his ordinary experiences to either an abstract

34. Slosson, "That Elusive Fourth Dimension," p. 275.

conception or an experiential conception of a fourth dimension. If the abstract conception was to be understood at all, it had to be in abstract terms. The common experiences of men were not the starting point for understanding the basic concepts or the extraordinary experiences of relativity theory. This fundamental limitation in the analogical method used by popular science destroyed its attempt to make the theories of relativity comprehensible in terms of common experience to the layman. The attempts to explain relativity in terms of common experiences led to the famous paradoxes of relativity which were, by their very nature, incomprehensible.

Relativity theory had changed the character of the science of physics. At one stroke, physical theory was removed from the world of common experience and common sense. But more than this, the theories of relativity made it clear that physics and persons have no contact with a noumenal reality.

Physics was limited to the measured fact. It could make no statements about a world behind the measured fact. As the French scientist-philosopher Gaston Bachelard has remarked, according to the theory of relativity, the reality men confront is a verified reality, not a given reality. "The objective world is the aggregate of facts verified by the science of our time." The Einsteinian editor for *Scientific American*, J. Malcolm Bird, described this phenomenalism in 1922: "We must put away as metaphysical everything that smacks of a 'reality' partly concealed behind our observations. We must focus attention upon the reports of our senses and the instruments that supplement them."[35] Phenome-

---

35. Gaston Bachelard, "The Philosophic Dialectic of the Concepts of Relativity," in *Albert Einstein: Philosopher-Scientist*, ed. Paul Arthur Schlipp (1949; reprint ed., New York: Harper & Row, 1959), II, pp. 570–71. This point is reinforced in Hans Reichenbach, "The Philosophical Significance of the Theory of Relativity," in ibid., I, p. 291. J. Malcolm Bird, ed., *Einstein's Theories of Relativity and Gravitation: A Selection of Material from the Essays Submitted in the Competition for the Eugene Higgins Prize of $5,000* (New York: Scientific American Publishing Co., 1922), p. 24. Also see Robert D. Carmichael et al., *A Debate on the Theory of Relativity* (Chicago: Open Court Publishing Co., 1927), pp. 145–46; Russell, *The A B C of Relativity*, pp. 226–27; Eddington, *Space, Time and Gravitation*, p. 185; Herbert Dingle, *Relativity for All* (London: Methuen, 1922), p. 67. The English scientists and philosophers tended to dwell on this problem more than their American counterparts because they had to deal with more extreme forms of empirical philosophy.

Also see Lotka, "A New Conception of the Universe," p. 485; Stewart, "The Nature of

nalism resulted from two consequences of the relativity theory: first, relativity theory had predicted and explained natural phenomena which had been impossible to predict on the basis of previous experience alone and which were unavailable to native or given experience; second, relativity theory made men dependent on the manipulation of instruments and experiments to reveal the natural world that cannot be directly experienced.

The operational interpretation of phenomenalism, which was implied in the second consequence of relativity theory, was brilliantly set forth by P. W. Bridgman, professor of physics at Harvard, in *The Logic of Modern Physics* (1927), a work well received by the lay press. Although Bridgman was hostile to the theories of relativity because he did not think they gave explanations for phenomena, his discussion of the revolution caused by relativity theory for the comman man as well as for the physicist best summarized the inadequacy of common experience.

Einstein's theories determined that reality cannot be described in terms of naive (given) experience. Reality must be described in terms of the operations or manipulations by which reality is measured. The naive experience of absolute simultaneity cannot be duplicated or verified by experiment; therefore, absolute simultaneity cannot be a characteristic of the universe. The naive experience of absolute physical dimensions cannot be detected by experiment; therefore, it cannot be a characteristic of the physical universe. All attempts to measure the absolute dimensions of bodies fail because of consequences implicit in the operation of establishing congruence when the speed of light is finite and constant.[36] The operational character of science explained why naive experience should not coincide with physical theory. Naive experience itself is an operation of measurement but with very limited equipment like eyes. Physicists can utilize far more per-

---

Things," p. 10; Claude Bragdon, "New Concepts of Time and Space," *Dial* 68 (February 1920): 187; Sydney T. Skidmore, "The Mistakes of Dr. Einstein," *Forum* 66 (August 1921): 130–31.

36. Alfred North Whitehead has a discussion of the establishment of congruity more critical of Einstein than the discussion by Bridgman. See Whitehead, "An Enquiry Concerning the Principles of Natural Knowledge," ch. 4, in *Alfred North Whitehead: An Anthology*, ed. F. S. C. Northrup and Mason W. Gross (New York: Macmillan, 1961).

ceptive instruments. Because the operations and accuracy of measurement in physics are different from those of the sensory faculties, the "experience" of the former will be different.[37] Naive experience is inadequate to probe reality and to understand physical theory because its measuring equipment is too crude, is limited to a relatively stable earth, and is unable to perceive directly more than two dimensions of space or to perceive the high velocity of light.

The character of modern science had established an impassable barrier between the layman and the physicist. To the extent that the popularizers of relativity theory attempted to remove that barrier, either they made the new physics apparently lucid by distorting it or they brought confusion to the lay mind by explaining it in the wrong terms.[38]

The Inadequacy of Common Sense

As confusing to the layman as the inadequacy of his common experience to understand the theories and physical reality was the inadequacy of his common sense. As Bertrand Russell observed, "The theory of relativity depends, to a considerable extent, upon getting rid of notions which are useful in ordinary life."[39]

Common sense does not refer to the Scottish philosophy of common sense, which was primarily an Enlightenment philosophy. Although it had great influence in American philosophy through the first half of the nineteenth century, it was shattered in the second half of the century by the rise of the functional psychology of William James and in the twentieth century by the instrumental philosophy of John Dewey. By *common sense*, rather, is meant the concepts formed on the basis of naive experience,

37. P. W. Bridgman, *The Logic of Modern Physics* (1927; reprint ed., New York: Macmillan, 1961), pp. 4–5, 23.

38. I have deliberately chosen not to discuss the question of whether experience of any category, even the experience of the physicist, can yield knowledge of the character of reality. I have considered to be pertinent only the discussion of the manner in which common experience was inadequate to understand relativity theory and the reality it describes.

39. Russell, *The A B C of Relativity*, p. 5.

such as time, space, and force. The invalidation of these concepts followed from the invalidation of common experience and was the second aspect of the reputed incomprehensibility of the theories of relativity.

The paradoxes of relativity forced the public to realize the inadequacy of their common sense. Most of these paradoxes involved the relativity of time and were based on the impossibility of absolute simultaneity. These paradoxes are quite familiar, for example, the paradox of the twins. One twin remains on earth. The other twin is orbited around the earth at a very high velocity. Because of the relativity phenomenon of the slowing of time rates with high speeds, when the orbiting twin is returned to earth, he discovers he has aged less than his brother. Another paradox involves a man who travels at nearly the velocity of light to a distant star. When he arrives he discovers that he has not aged nearly as much as the persons remaining on earth. All these paradoxes would occur if clocks instead of persons were the subjects. The paradoxical character of these illustrations of relativity phenomena arose not simply from the unfamiliarity of the phenomena but from the attempt to explain and to understand the theories in the old and inadequate common concepts. There were no paradoxes in the new physics itself. Though these paradoxes occasionally evoked humor, the general public reaction was unrelieved bewilderment.[40]

Books and articles analyzing the theories in some depth contained the realization that altered experience of the world demanded altered concepts of the world. But since the new

40. For a full elaboration of the paradoxes, see Edwin E. Slosson, *Easy Lessons in Einstein: A Discussion of the More Intelligible Features of the Theory of Relativity* (New York: Harcourt, Brace and Howe, 1920), pp. 15–17. Eddington, *Space, Time and Gravitation*, p. 27. "Repudiation of Common Sense by the New Physics: Has the Paradox of Matter and Motion Been Carried Too Far?" *Current Opinion* 64 (June 1918): 406–07; "Changing the Mind Gears," *Literary Digest* 64 (January 24, 1920): 29; "Einstein's Finite Universe [editorial]," *Scientific American* 124 (March 12, 1921): 202; Alexander McAdie, "Relativity and the Absurdities of Alice," *Atlantic Monthly* 127 (June 1921): 811–14. Humorous treatments of the paradoxes are provided by Robert Benchley, "All About Relativity: Einstein's Theory for the Lay-Mind in Simple Terms," *Vanity Fair* 13 (March 1920): 61; "In Darkest Einstein," *Nation* 105 (April 17, 1920): 503; Charles Martin, "How to Prove the Einstein Theory with the Aid of a Motor [Car]," *Vanity Fair* 21 (October 1923): 79; Ralph Barton, "Relativity," ibid. 21 (April 1924): 45.

experiences were unavailable to the layman, the new concepts (except in totally abstract forms) were, too. Accepting Bridgman's operationalism in its strongest interpretation, it can be understood that because the operations, whether mental or experimental, which established the new physics were unavailable to the layman, the new concepts were also. This, however, injects an emphasis into the lay response that may not have been conscious in the 1920s. At the time, the repudiation of common sense was synonymous with the repudiation of concepts which had common experiential references and their replacement by abstract concepts which had none. In common sense as in classical physics, a physical concept had been a mental image. Lorentz, for example, conceived of the phenomenon of contraction of dimensions in terms of a change in the electrostatic forces between atoms. These forces changed because of the passage of the object through the ether. The concept of forces, of course, had experiential references. Force (and contraction) were anthropomorphic qualities derived from the experience of muscular contraction and external resistance to muscular contraction.[41] In relativity theory, however, the contraction of dimensions lost this physical peculiarity. Instead, contraction appeared as a natural consequence of the complexities of description in a world in which absolute motion cannot be detected. Relativity theory postulated no mechanisms behind the phenomena of nature—no little muscles pulling together the atoms of an object.[42] The relativity concept of contraction is wholly abstract, without any anthropomorphic qualities which the layman could understand. The relativity concept

41. This classical concept of force was well stated by William Francis Magie, who opposed relativity theory, in his textbook, *Principles of Physics* (New York: Century, 1911), p. 9. "Man possesses a sense, called the motor sense, by which he is conscious of his exertion of *force*. He finds, by common experience, that by the exertion of force he may move a body, either altogether out of its place, or so as to distort it, or he may simply keep it at rest in some position from which it would move if he were to cease exerting force. From experiences of this last sort especially he has inferred that the body which he keeps at rest is exerting on him a force or *reaction* equal and oppositely directly to the force which he is exerting, and is at rest because of a balance between the forces. He thus infers that bodies exert forces on each other, and when he perceives results taking place without his agency, which he could have produced by exerting force, he ascribes them to forces exerted by other bodies."

42. See Eddington on this, *Space, Time and Gravitation*, pp. 17–29, esp. p. 28.

could not be imagined because there was nothing to be felt or to be imaged.[43]

The layman would inquire of the physicist: "I understand when you say that the contraction of dimensions is a consequence of the attempt to describe an object moving relatively to ourselves. But what happens to the *real* shape and *real* size of a moving body?" Given this question, the physicist's answer could only further confuse the layman: "Nothing happens, because the body does not have a *real* shape and a *real* size!"[44] The layman's common concept of reality was misleading him as to the meaning of the theories of relativity. To explain this, the relativist had to enter a discussion of epistemology.

The empiricism to which the popularizers and interpreters of relativity adhered asserted that concepts were derived entirely from perceptions of the world.[45] Edwin E. Slosson explained that man's mental world (which presumably included the mental world of relativist physicists) was a synthesis, depending primarily on contemporaneous (or simultaneous) association of sensations in the brain. Thus the concept of a box is derived, in general, from the perception of many individual boxes. The perception of any box in particular depends on the contemporaneous association of visual perception of the sides and visual inference of its volume, the smell of the glue and stain, the texture of the cardboard, the

43. This was the meaning behind the *Literary Digest*'s apt phrase, "changing the mind gears," in the article of that title, p. 29. See also Lotka, "A New Conception of the Universe," p. 482. The *Scientific American* discussed a similar dilemma with regard to the attempt to imagine the concept of an unbounded, yet finite, universe, in the editorial "Einstein's Finite Universe," p. 202.

44. This was well said by Dr. Morton C. Mott-Smith, of Los Angeles, in "Relativity and the Layman," in J. Malcolm Bird, ed., *Einstein's Theories of Relativity and Gravitation*, p. 106.

45. For example, Edwin E. Slosson, "How We Make Our Mental World," *Keeping Up With Science* (New York: Blue Ribbon Books, 1924), pp. 153-57; Bird, ed., *Einstein's Theories of Relativity and Gravitation*, pp. 72-73.

It is not necessary to discuss the well-known revival of empiricism in England led by Bertrand Russell, or of the Vienna school of positivism, to demonstrate that empiricism was the dominant epistemology of the 1920s. My concern is only to show that in the task of popularizing the theories of relativity, the empiricism led to a situation in which the common man (and all men in a strictly empirical view) was prevented from having direct acquaintance with reality, and, furthermore, that the common man was prevented from having an acquaintance with the concepts that gave the physicist his knowledge of physical reality.

sound the box makes when it is scraped with a finger, and its dull brown color. If these qualities were not contemporaneously associated, the concept of a box would be different. If the olfactory senses were impaired such that a smell sensation required a day to reach the brain, the conception of a box would be either something that leaves an odor that is perceived when it is not necessarily present or the simultaneity of the odor and the presence of the box would have to be inferred.

The destruction of the concept of absolute simultaneity by relativity theory destroyed the concept of objectively contemporaneous events. Therefore, the scientific description of the world could not be reduced to the logical and descriptive simplicity of the example above of the box. For this reason, the attempt of the popularizers to describe scientific empiricism in terms of personal empiricism had to be misleading. There was no strictly logical analogy. Here was another barrier to explaining the theories.

The basis for the simple empirical origin of concepts was undermined, second, because certain scientific concepts like *atom* could not be reduced directly to sensory statements such as *atom now* (I see an atom now). Many scientific concepts were constructs whose elements had no direct reference to the "object" or the "event" the construct signified. They might, as Bridgman insisted, for example, have reference only to operations like determining congruence. There were, in other words, no concepts in the new physics analogous to the concepts of common sense, which were concepts derived in a direct manner from the contemporaneousness and unity of common qualitative experience.

The layman's experience and concepts were, therefore, equally separated from the experience and concepts of the physicist and from physical reality. Common experience and common sense apparently gave knowledge of the qualitative character of nature, but the theories of relativity did not. In physical theory, words like *motion, space,* and *time* designated the observer's relationship to nature, but not nature itself. Defined mathematically, such concepts as *atom* were symbols designating sequences of phenomena. The concepts of physics gave no clue to the character of nature, they only prescribed the operations by which men's minds could accurately run "parallel" to the processes of nature.[46] But the concepts of common sense were inadequate to do even this.

Antithetical Realms of Being

The impact of the theories of relativity on the consensus for science was summarized in the new perspective the theories provided on the philosophy of John Dewey. Dewey's instrumental interpretation of ideas underlay the argument of Slosson that scientific ideas were the chief causes of cultural change. Popular science attempted to overcome the alienation of the layman from science by imparting scientific values to him and relating these values to other cultural values. This attempt to impart the values of the scientist to the layman was predicated on the assumptions that the scientific method was only a more refined version of the common man's method of thinking and that scientific objects (atoms, photons, and so on) were of the same kind as objects of common sense (shoes, cabbages). These were also the fundamental assumptions of John Dewey's philosophy. Dewey stated this quite bluntly: "There is no difference of kind between the methods of science and those of the plain man." The method of the scientist and of the plain man differed only in the problems they attempted to solve, not in their stages of thought or in their logic. Neither did the objects of science conflict with the objects of common sense. Ernest Nagel has succinctly stated Dewey's thesis on this: " 'The central thesis of Dewey's theory of science is that it does not disclose realms of being antithetical to the familiar things of life.' "[47]

46. J. Malcolm Bird had a satisfactory summary of this in *Einstein's Theories of Relativity and Gravitation*, chs. 2, 3, and 4.

47. Dewey, *Essays in Experimental Logic* (Chicago: University of Chicago Press, 1961), p. 86. Ernest Nagel, "Dewey's Theory of Natural Science," in *John Dewey: Philosopher of Science and Freedom*, ed. Sidney Hook (New York: Dial, 1950), quoted in Gail Kennedy, "Science and the Transformation of Common Sense: The Basic Problem of Dewey's Philosophy," *Journal of Philosophy* 51 (May 27, 1954): 318. Gail Kennedy's article is a defense of Dewey's philosophy against Ernest Nagel's attack on it. In chapter 4 of this work, I am on Nagel's side, arguing that science did not unify the scientist and the common man. I am, nevertheless, indebted to Gail Kennedy's article for the lucid manner in which she has formulated the problem of common sense.

It was, however, precisely a realm of being antithetical to the familiar things of life which the theories of relativity concerned. It was a realm so unfamiliar and so opposed to common experience that what it was could not even be communicated lucidly in concrete terms. High velocities of light, relative time, physical dimensions contracting according to high speeds, four-dimensional space-time were all scientific objects which conflicted with objects of common sense. The scientific method used by the scientists was so completely removed from the picture of the scientific method sketched in popular science as to contradict it. Slosson said that the scientific method was primarily factual discovery and secondarily a matter of grand visions. But Einstein's theories were primarily grand visions. Factual discoveries derived from them and came after them.

The motivations and goals of the theorists of relativity were not what Dewey and Slosson had thought. Dewey said the purpose of thought was the control of the environment for man's progress. Slosson had emphasized the hard work, the dirty hands, and the social goals of the researcher. But Einstein did not say that the purpose of his theories was to change the content of the daily life of men. The purpose was "to arrive at those universal elementary laws from which the cosmos can be built up by pure deduction." For Einstein, even the motivation to engage in theoretical physics was an attempt to escape from the "noisy cramped world" of personal and temporal problems.[48] It should not be surprising that the theories of relativity and the controversy over them were detrimental to the consensus on science.

48. Albert Einstein, "Principles of Research" (Address, 1918), in *Essays in Science* (New York: Wisdom Library, 1934), pp. 2, 4.

(5) IDEOLOGY: THE SCIENTIFIC BASIS OF PROGRESS

The concept of progress has always been attended by the opposite concept of degeneration. The concept of progress originated in the seventeenth century in opposition to the notion prevalent among classicists that nature's powers had become diminished since the ancient world and that she was "no longer capable of producing men equal in brains and vigour to those whom she once produced."[1] On the literary side of this quarrel, the modernists, who claimed that literary geniuses as great as the ancient geniuses were still being produced, triumphed over the classicists with the aid of Newtonian science which presumed that the processes and forces of nature were consistent throughout time. This scientific concept of uniformitarianism, extended in the nineteenth century to include geological and biological processes, provided a stable platform on which the belief in progress could be erected. Uniformitarianism allowed the indefinite time span required for the progressive accumulation of knowledge and improvement of man's moral character and culture.

What science gave to the belief in progress, however, it also threatened to take away. The three laws of thermodynamics, formulated in the nineteenth century, apparently indicated that the degeneration of nature was a basic feature of physical reality. The second law, established by the German physicist Rudolph Clausius in 1850 and in a more general form by William Thom-

---

1. J. B. Bury, *The Idea of Progress: An Inquiry into Its Origin and Growth* (1932; reprint ed., New York: Dover Publications, 1955), p. 79. See also Richard Foster Jones, *Ancients and Moderns: A Study of the Rise of the Scientific Movement in Seventeenth-Century England* (1936; reprint ed., Berkeley: University of California Press, 1965).

son, Lord Kelvin, in 1851, stated that the amount of energy unavailable for work in a closed thermodynamic system (entropy) always increased. According to the third law, the increase of this amount approached a maximum at which point no energy would be available for work.[2] When these two laws were extrapolated from the empirical systems for which they had been observed to be valid to the universe itself considered as a closed system, they predicted a final state of universal inert uniformity, all bodies having the same temperature, no energy available for work, and all change, including life, ended. This was the heat death.

Several criticisms of the cosmic extrapolation from the second and third laws were raised. In his theory of heat (1875), the English physicist, James Clerk Maxwell, demonstrated that the intervention of intelligence could suspend the second law. More important, the interpretation of the science of thermodynamics in terms of statistical mechanics by Ludwig Boltzmann and J. Willard Gibbs reduced the increase of entropy to a probability, thus denying the necessitarian character of the universal heat death. It was not totally improbable that the universe should return to a situation of high potential energy.[3]

The new physics, beginning with Roentgen's discovery of X rays in December 1895 and Becquerel's discovery of natural radioactivity the following year, seemed, however, to give new direct evidence for the validity of the second law. Certainly the discoveries of these scientists gave impetus to cosmological speculation concerning the second and third laws. The natural dissipation of available energy as radiation without a concomitant creation of energy implied that "the universe is losing its available energy and is going steadily to a condition of rest and extinction." For this reason, the Curies' discovery of radium was, as

2. The relevant papers of Clausius and Thomson are reprinted in William F. Magie, ed., *The Second Law of Thermodynamics: Memoirs by Carnot, Clausius, and Thomson* (New York: Harper & Brothers, 1899). For a discussion of these laws, see P. W. Bridgman, *The Nature of Thermodynamics* (New York: Harper, 1961).

3. A bibliography of the scientific literature of thermodynamics to 1890 is Alfred Tuckerman, *Index to the Literature of Thermodynamics* (Washington, D.C.: Smithsonian Institution, 1890); see pp. 96–98 for works on the second law.

Henry Adams said in his autobiography, a "metaphysical bomb."[4]

Einstein's special theory of relativity, moreover, gave indirect evidence for the validity of the second law. The special theory established a temporal and directional order to physical events, leaving entropy as an invariant, with at least mathematical reality. That is to say, the special theory established that the order of physical events was irreversible which was of course what the second law of thermodynamics asserted. The implication of a directional order for physical events sprang from the special theory's denial of the universal validity of physical simultaneity. There was no "state of the universe as a whole" at any particular moment which meant that the universe could never be exactly the same at any two moments separated by a causal action. Thus a cyclically changing cosmos was denied and a "linear" cosmic evolution established.[5]

In the decade following the First World War, the English astrophysicist, James H. Jeans, became a leading proponent of the cosmological validity of the second law of thermodynamics and the notion of the heat death.[6] Jeans's popular writings made

4. Robert Kennedy Duncan, *The New Knowledge: A Popular Account of the New Physics and the New Chemistry in Their Relation to the New Theory of Matter* (New York: A. S. Barnes, 1905), p. 241. Duncan's book, which Duncan wrote after S. S. McClure asked him to write a story on science, was one of the earliest American popularizations of the new physics. Duncan himself doubted the universal validity of the second law, but admittedly had no evidence denying its universal validity. Henry Adams, *The Education of Henry Adams: An Autobiography* (Boston: Houghton Mifflin, 1927), p. 452.

5. The positivist philosopher Hans Reichenbach had formulated a causal theory of time based on the special theory, which dealt with the directional character of change, as early as 1921. This theory is discussed in his *The Direction of Time*, ed. Maria Reichenbach (Berkeley: University of California Press, 1956), pp. 24–27. Reichenbach discusses the irreversibility of time direction, pp. 108 ff. See also Morris R. Cohen, *Reason and Nature: An Essay on the Meaning of Scientific Method* (1931; reprint ed., London: Collier-Macmillan, 1964), pp. 235–36; Richard C. Tolman, *Relativity, Thermodynamics, and Cosmology* (Oxford: Clarendon Press, 1934), pp. 131–36. A more recent discussion of the directional character of change is provided by Milič Čapek, "The Theory of Eternal Recurrence in Modern Philosophy, with Special Reference to C. S. Peirce," *Journal of Philosophy* 57 (April 28, 1960): 289–96.

6. James H. Jeans, *Astronomy and Cosmology* (1928; reprint ed., Cambridge, England: University Press, 1929), ch. 4; and *The Mysterious Universe* (New York: Macmillan, 1930), ch. 3, esp. pp. 79–80, 154. It is not out of place here to note that the third law of thermodynamics, that entropy approaches a maximum (the heat death), is not a logically necessary inference from the second and first laws of thermodynamics. There are

this cosmological question of enormous public interest, justly reflecting the scientific importance of the question.

The revived pessimistic doctrine of the second and third laws of thermodynamics had great consequences for an optimistic and progressive ideology of science. The indefinite time span required for the progressive improvement of man's character and culture would be implicitly denied if natural processes were to end at any particular moment, however remote in the future. This threat was especially important because the theory of evolution—which was assimilated into the concept of progress—required great periods of time. Whereas the European concept of progress was millennial, foreseeing the end of historical time and progress in a secular utopia, the American concept of progress was open-ended.[7] The impact of the second law on the concept of progress was revealed in that careful mind of Henry Adams who applied this law of physics in analogy to the utilization of energy in history to predict the end of civilization.[8] So Joseph Conrad would cry, " 'If you believe in improvement you must weep.' "[9] And in H. G. Wells's novel *Tono-Bungay* (1909), radioactivity became the symbol of decay, the " 'ultimate eating-away and dry-rotting and dispersal of all our world.' "[10]

---

several possible inferences from the second law, one of which is that entropy should always increase, never reaching a maximum. This latter inference is not only logically possible, but development of the general theory of relativity indicated it was physically possible. The relativistic physicist, Richard Tolman, in 1934 published a theory based on the general theory of relativity to this effect (Tolman, *Relativity, Thermodynamics, and Cosmology*, pp. 326–30, 439–44). I have found no echo of this part of Tolman's work in Robert Millikan, the major figure in this chapter, in the 1920s, even though Tolman taught at the California Institute of Technology which Millikan headed. Therefore, I have not developed this topic in the text.

7. This is the interpretation of Rush Welter, "The Idea of Progress in America: An Essay in Ideas and Method," *Journal of the History of Ideas* 16 (June 1955): 403–05. The utopian element was never lacking entirely in the American concept of progress. Particularly in the twentieth century, engineering gave impetus to the concept of technological utopia.

8. Henry Adams, "A Letter to the American Teachers of History," in *The Degradation of the Democratic Dogma*, ed. Brooks Adams (New York: Macmillan, 1919), pp. 137–266. See also the discussion of William H. Jordy, *Henry Adams: Scientific Historian* (New Haven: Yale University Press, 1952), ch. 6.

9. Quoted in Jerome Hamilton Buckley, *The Triumph of Time: A Study of the Victorian Concepts of Time, History, Progress, and Decadence* (Cambridge, Mass.: Harvard University Press, Belknap Press, 1966), p. 67.

10. Quoted in ibid., p. 88.

The pessimistic interpretation of the second law and heat death concept led to another consequence: the implicit denial of the progressive character of the scientific method. Progress due to the scientific method was dependent not only on the method's technique for the manipulation of phenomena but also on its relationship to nature. If the physical reality on which the scientific method depended was in decay, the method itself could not produce progress merely by revelation of nature. In the 1920s, Robert Millikan, the leading spokesman for the national scientists, attempted to refute the second and third laws of thermodynamics and to discover a basis in nature for the progressive character of the scientific method. Millikan clearly hoped that nature itself would guarantee the progress which was the central value of the ideology of national science.

Robert Millikan's Search for Cosmic Progress

The discovery of natural radioactivity in 1896 immediately raised a profound question: What is the character of matter? Before the discovery of radioactivity, it was generally assumed that elemental permanence was the essential characteristic of matter. Radioactivity, however, reintroduced the Heraclitean notion that qualitative change was the essential characteristic. The atoms of the heavier elements, at least, disintegrated into simpler atoms, accompanied by the emission of particles. Radioactivity seemed to be as basic a property of matter as mass. In classical physics, qualitative change had been an illusion of the senses because the only "change" was the movement of particles from position to position. But radioactivity, as Millikan said, gave strong evidence that "the nature of the atom itself" changes.[11]

Understanding how this change in the nature of the atom occurred has been, of course, the great task of modern physics. But the discovery of radioactivity set out other important tasks as well. One was understanding the source of subatomic energy.

11. Robert A. Millikan, "Recent Discoveries in Radiation and Their Significance," *Popular Science Monthly* 64 (April 1904): 492.

Radioactivity indicated the existence of a huge quantity of such energy. No known physical or chemical processes were capable of yielding energy for 36,000 to 900,000 years (Millikan's estimates of the life span of radium). Millikan doubted that this subatomic energy would have any practical use (except, perhaps, in treatment of cancer), because only "three substances . . . are disengaging it and these are changing so slowly that the rate of evolution of energy is almost infinitesimal."[12] But in 1904, he could have had no intimation of the horrifying adventure through which atomic energy was to lead twentieth-century man.

Another great task assigned by the discovery of radioactivity concerned what was not discovered. The disintegration of the complex atoms into simpler atoms suggested the ancient alchemical dream of transmuting the base metals into precious metals. But the discovery of radioactivity was of the transmutation of the precious into the common, of uranium into lead. There was no discovery of the creation of the heavy elements out of the lighter elements. Not only did matter appear to be unstable, but it appeared to be in decay, a decay perhaps originating with the universe itself. Energy put into matter at creation was irretrievably leaking away. The second law of thermodynamics seemed ultimately confirmed.

The second law of thermodynamics could be invalidated only if atoms were being synthesized as well as disintegrated into radiation. Millikan searched for the possibility of synthesis in 1904, early in his career, by drawing an analogy between life processes of change and atomic processes of change. All organic compounds were complex molecular structures created out of simpler structures by the processes of life. Why should not there be, he asked, some similar inorganic agency synthesizing heavy, complex elements out of the lighter, simple elements. "It would be rash to attempt to give any positive answer to such a query," Millikan went on, "yet the fact that radium now exists on the earth, taken in connection with the fact that the life of radium is short in comparison with the ages in which the earth has been in

12. Ibid., p. 499.

existence, certainly seems to point to an affirmative answer." The analogy between the organic and the inorganic forces of change was so powerful in Millikan's mind that later he would describe radium as "alive," meaning that it had life's characteristic of qualitative change. "Dead matter is very much alive in a new and undreamed of way." In the 1920s, with apparent evidence that most elements, not only the heavier elements, were or could be radioactive, all of nature became alive for Millikan. As late as 1923, however, discovery of the existence of the synthesis of heavy, complex elements out of the lighter, simple elements still eluded science. Yet, he did not doubt the discovery would be made.[13]

Millikan himself was to make the "discovery" of atomic synthesis in the mid-1920s. This discovery came in his research on the cosmic rays. He was to propose that cosmic rays, which could be studied empirically, were given off in the synthesis of the elements, which could not be so studied. Millikan's interest in cosmic rays dated back to 1914 and 1915, when he and students in his seminars discussed recent European experiments on the detection of rays in the atmosphere. These experiments had been designed to determine the cause of common electroscopic discharge. It had been observed that an ordinary, charged electroscope, even when well insulated by a lead shield, would have a slow discharge. The suggestion was made that the discharge was due to highly penetrating rays of a radioactive origin on earth. To test this hypothesis, electroscopes were sent aloft in balloons. If the rays originated from the earth, the electroscopes should not discharge or should discharge at a slower rate in the upper atmosphere. The prewar balloon experiments, however, revealed that the rate of discharge increased in the atmosphere. The new hypothesis was made that the rays discharging the electroscope came from above the earth's surface. At this point in the research the war intervened and no more research was undertaken until the postwar decade.

13. Ibid., pp. 498–99. Millikan, "The New Physics," *School Review* 23 (November 1915): 614–15. Millikan, "Gulliver's Travels in Science," *Scribner's Magazine* 74 (November 1923): 584.

During the war, Millikan's meteorological research for the army fostered his interest in balloon experimentation and cosmic rays. He decided after the war to take up research where the Europeans had left off. Cosmic rays became one of his major research efforts during the entire interwar period, producing many doctoral projects for his graduate students.[14] Beginning in 1922 (shortly after his arrival at the California Institute of Technology), Millikan, together with colleagues, undertook two types of cosmic ray experiments: balloon experiments, in which electroscopes were carried into the upper atmosphere to determine rates of discharge and to locate the origin of the rays; and penetration experiments, using lead on Pikes Peak and electroscopes underwater in high mountain lakes, to determine the penetrating power and frequency of the rays. The latter experiments would allow the deduction of the amount of energy necessary to produce the rays. One early result of these experiments was that radiation with properties described by the prewar investigators did not exist. Instead, two types of radiation were discovered. One was a hard and highly penetrating radiation; the other was a secondary, soft, and less penetrating radiation. These radiations had their origin, at least, in the atmosphere above the highest mountains and it was presumed the origin was cosmic.

Although later researchers considered Millikan's early experiments of dubious value, the value of the experiments was far less dubious than the value of the speculations Millikan based on them. There were only two atomic processes which Millikan could think of that yielded the enormous energy to create rays with the penetration power of hard cosmic rays. The first process would have been an atomic synthesis of helium out of hydrogen. The second would have been the capture of a free electron by an atomic nucleus. For reasons not entirely clear, in 1925 Millikan dismissed the first alternative, basing his theory of the origin of cosmic rays on the second process. Electron capture would release twelve million volts of energy at a minimum, enough to produce

14. The foregoing account of the cosmic ray research is based on Robert A. Millikan, *Autobiography* (New York: Prentice-Hall, 1950), pp. 146–47, 209–11.

the lower order of the cosmic rays.[15] Within the next several years, however, Millikan rejected the electron-capture hypothesis and adopted the first alternative, the atom-synthesizing hypothesis. This latter theory had the virtue of explaining the higher magnitudes of energy which some cosmic rays demanded, as well as refuting the cosmological validity of the second law of thermodynamics.

Millikan's atom-building hypothesis was quite simple. The helium atom could theoretically be formed by the synthesis of four hydrogen atoms. The masses of these atoms were known with great accuracy because of the recent work of the English physicist, Francis W. Aston. Although the helium atom was composed of four hydrogen atoms, the mass of the helium was slightly less than four times the mass of the hydrogen atom. Therefore, when four hydrogen atoms fused to form a helium atom, some mass had to be dissipated as energy. This same process occurred for the formation of oxygen, silicon, and iron atoms from helium or hydrogen atoms. Using Einstein's equation for the equivalence of mass and energy, $E = Mc^2$, the magnitude of the released energy could be calculated. It turned out that the range of energy released from the series of atom syntheses from helium to iron more than sufficed to produce all known cosmic rays. This was all qualitative evidence, however, and did not prove the existence of such atom-building or that cosmic rays, as a matter of fact, were produced by atom-building. The theory indicated only that cosmic rays could be so produced. Millikan thought that the absorption coefficients of the rays (their penetration power) provided the direct quantitative evidence linking the rays to atom-building. The absorption coefficients of observed cosmic rays grouped into three classes—0.35, 0.08, 0.04 (per meter of water). The theoretical formation of helium, oxygen, and silicon out of hydrogen yielded energy able to produce penetration coefficients of, respectively, 0.35, 0.075, and 0.043. These theoretically derived coefficients correspond within the limits of

15. Millikan, "High Frequency Rays of Cosmic Origin," *Proceedings of the National Academy of Sciences* 12 (January 1926): 48–55.

error to the observed coefficients and provided the basis for a plausible argument that cosmic rays were produced by atom-building.[16]

The atom-building hypothesis was not a necessary hypothesis because it was not the only one consistent with the empirical evidence. James Jeans proposed, for example, the equally valid theory that cosmic rays were created by the annihilation of atoms—the collapse of orbital electrons into the atomic nucleus, resulting in their electrical neutralization and dissipation as energy. Millikan sought, therefore, additional evidence in favor of the atom-building hypothesis. He thought he had this evidence when he located the place of the cosmic rays' production. Since the annihilation of atoms was presumed to occur only inside stars, the discovery that the cosmic rays came to earth uniformly from all directions and at all times with equal intensity favored the atom-synthesizing hypothesis. If cosmic radiation came from the stars, the observed intensity of the radiation should vary with the attitude of the earth toward the sun and the major galaxies. Since this variance was not detected, Millikan concluded that the creation of the rays was occurring uniformly throughout interstellar space. This conclusion was reinforced by the improbability that atom-building occurred in the stars.[17]

For Millikan the importance of the question of the location of the production of cosmic rays went beyond the cosmic rays themselves. To locate the origin of the rays in interstellar space was crucial evidence for Millikan's theory of the cosmos. Even while he retained the electron-capture hypothesis, he was certain that the electron-capture formation of the rays had to be occurring in interstellar space.[18] The importance of interstellar space can be understood only in the context of his theory of the cosmos. This

16. Millikan and G. Harvey Cameron, "Evidence for the Continuous Creation of the Common Elements out of Positive and Negative Electrons," *Proceedings of the National Academy of Sciences* 14 (June 1928): 445–50.

17. Jeans, *The Mysterious Universe*, pp. 76–80. Millikan and G. Harvey Cameron, "Evidence That the Cosmic Rays Originate in Interstellar Space," *Proceedings of the National Academy of Sciences* 14 (August 1928): 637–41.

18. Millikan, "High Frequency Rays of Cosmic Origin," p. 54.

theory was admittedly indebted to the cosmic theory and concept of the energy cycle of a former colleague at the University of Chicago, William Duncan MacMillan, professor of astronomy.

MacMillan's theory of the cosmos and his theoretical cycle of energy attempted to reinstate the classical, Newtonian cosmology which had been replaced by a cosmology based on the general theory of relativity and to invalidate the second and third laws of thermodynamics for the universe as a whole. MacMillan hoped to widen the cosmological time span to accommodate the temporal requirements of evolutionary biology and geology and to undermine the cosmic pessimism of the second and third laws.

MacMillan thought it was possible to construct a rational cosmology based on some thirteen Newtonian postulates consistent with known optical evidence of the structure of the universe (radio sources not having been discovered). MacMillan's universe had Euclidean metric properties, was unbounded with a double-ended infinitude. Time was absolute, nondimensional, and independent of space.[19] MacMillan's cosmological theory was addressed to two questions: What became of all the energy (visible light, and other radiations) emitted by the stars? And, what is the source of subatomic energy? With regard to the first question, if all the energy emitted by the stars traveled uninterrupted through space, then the night sky should have been as bright as day. Since the night sky was not, the conclusion followed that some light was absorbed into space before reaching the earth, leaving the night dark and cold. MacMillan dismissed the hypothesis that invisible dark bodies absorbed this energy because the absorption of the energy would heat the bodies to incandescence, producing a visible glow, which should therefore be seen, but was not. The only alternative was that the star energy was absorbed by the ether. (MacMillan did not enter into the question of the physical reality of the ether. The ether was for him at least only a hypothecated entity maintained for the pur-

19. William Duncan MacMillan, "The Structure of the Universe," *Science*, n.s. 52 (July 23, 1920): 67–74; and especially "Some Mathematical Aspects of Cosmology, II. Cosmology," *Science*, n.s. 62 (July 31, 1925): 97–99.

pose of the argument.) If absorbed by the ether, the lost star energy had to do work. What work could it do? MacMillan suggested that it built atoms. Bouncing around the ether, light particles would occasionally collide and hook together, MacMillan said, employing a Lucretian metaphor, to form rudimentary, probably hydrogen, atoms.

As these free atoms in interstellar space became locked in larger associations, the gaseous nebula would be formed. Eventually, the gravitation of the nebula would contract its volume, increasing the density, pressure, and temperature of the mass. When the pressure and temperature reached a critical point, the disintegration of atoms was forced, with the accompanying release of subatomic energy. This subatomic energy would further raise the temperature until the mass became a star, fueling on the disintegrating atoms. If the star lived only on its own atoms, it would quickly (in cosmological perspective) expire. This indeed was predicted by some classical theorists and was the reason for the cosmological time constriction which contradicted the historical evidence of geology and biological evolution for a long time span. MacMillan proposed, however, that there was no reason the stars should ever expire. In their passage through space, the stars should pick up enough free atoms to replenish their supply. "It is not necessary to suppose," MacMillan wrote, "that the universe as a whole has ever been or ever will be essentially different from what it is today."[20] The forms of the universe were permanent, even if individual representatives came into and passed out of existence.

MacMillan's theory of the energy cycle answered his second question concerning the source of subatomic energy. The key to this cycle was the equivalence of matter and energy, according to Einstein's special theory of relativity, a theory MacMillan ironi-

20. MacMillan, "Stellar Evolution: An Attempt to Correlate Two Outstanding Problems of Physics," *Scientific American Supplement* 87 (May 24, 1919): 336. See also "Cosmic Evolution, First Part: What Is the Source of Stellar Energies?" *Scientia* (Milan) 33 (January 1923): 3–12; and "Cosmic Evolution, Second Part: The Organization and Dissipation of Matter Through the Agency of Radiant Energy," ibid. (February 1923), pp. 103–12.

cally hoped to refute. This equivalence made possible a cosmic cycle of energy to matter to energy. This cycle was obviously outlined in the cosmology. Radiant energies from stars occasionally produced atoms, with the store of energy necessary to hold the atoms together. These atoms eventually coalesced into stars or were absorbed by stars, which, in turn, gave off radiant energies.

MacMillan had drawn a comforting vision. It described a universe which was Newtonian. The second law of thermodynamics, to which relativity theory gave indirect support, was not universally valid. Nature did not degenerate:

> The haunting fear of a general stellar death is gone and the forbidding picture of the galaxy as a dismal, dreary graveyard of dead stars fades away from our sight; and in its stead we see an indefinite continuation of our present active, living universe with its never-ending ebb and flow of energy.[21]

The least acceptable aspect of this vision was the conversion of free stellar energy into atoms in interstellar space. MacMillan's Lucretian suggestion of atoms bobbing around in the ether and snagging together by hooks was hardly convincing. It was Robert Millikan's theory of the creation of cosmic rays by atom synthesis which made MacMillan's suggestion scientific. Millikan's theory proceeded as follows: Atoms were disintegrated in stars under conditions of extreme heat and pressure. The electrons collapse into the nucleus, resulting in neutralization and the release of energy, as in Jeans's theory. It was to be expected that under exactly opposite conditions, free electrons and protons would coalesce into the simplest form of atom—hydrogen—and then condense into more complex atoms. In Millikan's theory, this condensation would release cosmic rays. Millikan could even borrow from the wave mechanics of quantum theory the notion that hydrogen atoms could "jump over a potential wall and find themselves together in a nucleus" as a helium atom, if the hydrogen atoms were free from conditions of pressure and heat. Abso-

21. MacMillan, "The Structure of the Universe," p. 73.

lutely zero temperature and zero pressure were precisely the characteristics of interstellar space.[22]

MacMillan and Millikan had described a steady-state cosmology. The universe was eternally running in the same cycle. This was opposed to other cosmologies which accepted the validity of the second law of thermodynamics and the relativistic denial of cyclical change. These latter cosmologies saw a beginning and end to the universe. It is not necessary to carry the story of the cosmic rays beyond this point, although the history of cosmic ray research is fascinating precisely because it leads to speculations such as Millikan's. None of Millikan's ideas about cosmic rays are now considered correct as he stated them. Cosmic rays are apparently produced by an atom-building process (first elaborated correctly by Hans Bethe), but this process occurs inside stars, not in interstellar space. Cosmic rays do not come to the earth equally from all directions and with equal intensity at all times, but where exactly they originate is not clear. Some version of the steady-state cosmology is held today by many leading astronomers, although not for the same reasons that it was held by Millikan.[23]

Why did Millikan hold so tenaciously to the steady-state, Newtonian cosmology and the invalidation of the second law of thermodynamics? One reason certainly was religious. Millikan believed in the classical doctrine of the immanence of the Deity in the universe.[24] His steady-state cosmos was more compatible

22. MacMillan, "Stellar Evolution," p. 322. Millikan, "The Evolution of the Universe," *Nature* 127 (October 24, 1931): 715. Millikan and Cameron, "Evidence That the Cosmic Rays Originate in Interstellar Space," pp. 639–40. See also Millikan, "Available Energy," in *Science and the New Civilization* (New York: Charles Scribner's Sons, 1930), pp. 96–109. MacMillan's endorsement of Millikan's cosmic ray theory was made in MacMillan, "The New Cosmology," *Scientific American* 134 (May 1926): 310–11.

23. The theories of MacMillan and Millikan, as well as other controversies over cosmic rays in the 1920s and early 1930s, are discussed in Harvey Lemon, *Cosmic Rays Thus Far* (New York: W. W. Norton, 1936). The recent state of cosmic ray research is reviewed with partisanship in Fred Hoyle, *Galaxies, Nuclei and Quasars* (New York: Harper & Row, 1965), chs. 3 and 4. Millikan's views changed after the great burst of research on cosmic rays in the 1930s. See Millikan, *Cosmic Rays* (Cambridge, England: University Press, 1939).

24. Millikan, "The Present Status of Theory and Experiment as to Atomic Disintegration and Atomic Synthesis," *Nature* 127 (January 31, 1931): 167.

with a doctrine of immanence than a cosmos which was in constant decay.

Millikan was also attracted to MacMillan's cosmology out of a scientific conservatism. Although the second law of thermodynamics was considered valid for local, closed systems, there was no direct evidence that it was valid for the universe as a whole. Millikan did not think that scientists could justifiably generalize from conditions on the earth to the entire universe. Radiation laws exhibited on earth would not necessarily hold elsewhere. The dogma of the second and third laws, Millikan argued, "rests squarely on the assumption that we, infinitesimal mites on a speck of a world, know all about how the universe behaves in all its parts."[25] This was, however, the subversion of a fundamental assumption of science, the universal validity of the laws of nature. It was unreasonable of Millikan to assert that the laws of nature might not hold throughout the cosmos because scientists had no evidence that natural laws were not universally valid. Millikan's ad hoc scientific agnosticism was misplaced. Scientifically, agnosticism was admirable when it was an admission that some questions had no scientific answers, but not when it denied to science the very tasks which were science's purpose. Millikan also found MacMillan's cosmology attractive out of a conservative desire to retain the Newtonian universe. MacMillan's rejection of relativity theory was straightforward. He maintained the scientific validity of "normal intuition" (common sense), which was decisively rejected by the special and general theories.[26] Although Millikan was prepared to use parts of relativity theory which suited his purposes, such as the equivalence of matter and energy, his reluctance to accept the relativistic cosmology had been evidenced as early as 1919, after the announcement of the results of the eclipse expedition.

Finally, there was the most important reason for Millikan's

25. Ibid.
26. MacMillan, "The Postulates of Normal Intuition: The First Speech of the Negative," and "The Fourth Doctrine of Science and Its Limitations," in Robert D. Carmichael et al., A Debate on the Theory of Relativity (Chicago: Open Court Publishing Co., 1927), pp. 39–63, 117–27. See also note 13, this chapter.

adherence to the steady-state cosmology and the invalidation of the second and third laws of thermodynamics: the ideological necessity of providing a *unified perspective* on nature, science, society, and progress. It was true that even if the second and third laws of thermodynamics were universally valid, the remoteness of the heat death would permit all practical human progress. The point was, however, that the daily degeneration of nature would not have permitted a scientifically optimistic and progressive attitude toward the universe.

The necessity for ideological unity was nowhere as well displayed as in Millikan's address, "Available Energy," delivered in September 1928, before the British Society of Chemical Industry. In this address, the cosmological basis for the progressive ideology of national science was explicitly laid out. One of the fears of the 1920s was that all the coal and petroleum reserves would be depleted within several decades.[27] This fear was important not only for the actual possibility of depletion, but because it threatened allegiance to the progressivism of the national science ideology. Depletion of the fuel reserves would be a local manifestation of the heat death. The recent industrialization of America seemed about to be throttled before it had the opportunity to diffuse economic democracy to all the people. Some engineers thought that the recently discovered subatomic energies could be utilized as fuel. But Millikan was forced to dismiss this possibility for theoretical reasons. Neither did artificial atomic fusion seem a possible source of energy since the interstellar conditions for atom synthesis could not be duplicated on earth. Process of elimination left only the sun as man's source of future energy. But was the sun a reliable source? Eddington and Jeans asserted that the sun was cannibalizing itself to provide radiation for man. It was a dying source of energy. Millikan's answer to this was, of course, the energy cycle: "When the matter of the sun has all been

27. Edwin Slosson was particularly concerned with the depletion of fuel resources. See Slosson, "Inventory of Energy," *Scientific Monthly* 16 (February 1923): 217–19, for an introduction to this subject. See also Slosson's papers in "International Research Council" file, box 9, Robert A. Millikan Papers, California Institute of Technology Archives (cited hereafter as Millikan Papers).

stoked into his furnaces and they are gone altogether out another sun will probably have been formed, so that on this earth or on some other earth—it matters not which some billions of years hence—the development of man may still be going on."[28]

Man's progress was assured not simply because science would provide answers to problems but because nature itself did not oppose man's progress. Scientists, tied by the scientific method to physical nature, could not provide answers if nature did not allow them. The unity of the scientific, progressive ideology was provided by the knowledge that the cosmos sanctioned this ideology. The ideology was not artificial; it followed necessarily from a scientific knowledge of physical reality. Man had "in all probability another billion years ahead of him, in which there is the possibility of his learning to live at least 'a million times more wisely' than he now lives."[29]

Millikan's assumption that nature favored progress was revealed least ambiguously in his refutation of the fear of the 1920s of some "diabolical scientist tinkering heedlessly, like the bad small boy, with enormous stores of subatomic energy, and some sad day touching off the fuse and blowing our comfortable little globe to smithereens."[30]

The fear that atomic energy might destroy the world had been expressed as early as 1915 by the British atomic physicist, Frederick Soddy. Before the First World War, Soddy had entertained typically nineteenth-century visions of the grand benefits for man to be drawn from the discovery of atomic energy. The world war shattered his belief in the beneficence of nature. "Imagine, if you can," Soddy requested his audience in a lecture of 1915, "what the present war would be like if such an explosive [utilizing atomic energy] had actually been discovered instead of being still in the keeping of the future." Such an atomic explosive, he prophesied, "is a discovery that conceivably might be made tomorrow, in time for its development and perfection for the use or destruction, let us say, of the next generation, and which, it is

28. Millikan, "Available Energy," in Science and the New Civilization, p. 113.
29. Ibid., p. 90.
30. Ibid., p. 95.

pretty certain, will be made by science sooner or later." Soddy's fears were echoed occasionally during the next decade. In September 1927, the bishop of Ripon (England) suggested that a ten-year moratorium be declared on scientific research. Physical science had been making so many discoveries and gaining such control over nature's forces that men had not been able to assimilate them or understand their implications. Until assimilation had been achieved, the bishop explained, "man cannot feel safe, and the very greatness of his recent achievements would seem to make his ruin more certain and more complete." Raymond B. Fosdick, a trustee of the Rockefeller Foundation and an acquaintance of Robert Millikan, also asked the question, in his book *The Old Savage in the New Civilization* (1929), whether science had presented man with achievements he could not use to his benefit because he could not control them.[31]

These reservations have not been presented because they were unusually prescient. These are concerns typical of an era of rapid scientific and technological change and could be found frequently in the Victorian era, for example. It was, rather, Millikan's reply to these charges that was important. The charges could not remain unanswered if the cosmological basis of the progressive ideology of science were to be preserved. For these charges carried the suspicion that nature itself had no bias in favor of man. To Soddy's fear that science would unloose atomic explosives on the world, Millikan replied that science's *only* important objective was "to find out the facts." Millikan apparently thought Soddy was accusing science of causing the recent world war. This was nonsense, Millikan argued, because war had been a habit of civilizations long before the rise of science. (This did not, however, refute the proposition that science caused this particular war.) Actually science was abolishing war by lessening

31. Frederick Soddy, *The Interpretation of Radium* (1909; reprint ed., London: John Murray, 1912), pp. 238–40. Soddy, "Physical Force—Man's Servant or His Master?" Address to the Aberdeen Branch of the Independent Labour Party, November 1915, *Science and Life: Aberdeen Addresses* (London: John Murray, 1920), p. 36. "Wants 10 Year Pause in Scientists' Efforts," *New York Times*, September 5, 1927, p. 3. Raymond B. Fosdick, *The Old Savage in the New Civilization* (Garden City, N.Y.: Doubleday, Doran, 1929).

its survival value. The fear of atomic explosives was unjustified because "every scientific advance finds ten times as many new, peaceful, constructive uses as it finds destructive ones." (Again, this did not refute Soddy's fear.) Millikan did, finally, address himself directly to Soddy's prophecy of atomic explosives. This was an unfounded apprehension, said Millikan, because "nature, or God, whichever term you prefer, was not unconscious of the wisdom of introducing a few fool-proof features into the machine [nature itself]." Science could prove that the supposed subatomic energy in the elements was "locked in." All the elements with the exception of the few naturally radioactive—uranium, radium, and thorium—were at their maximum stability. If their atoms were to be torn apart artificially, they would have no energy to give up. The fear of atomic explosives was a "hobgoblin," a "bugaboo," a "myth," created in ignorant minds. Nature did have, Millikan asserted, a bias in favor of man.[32]

The ten-year moratorium on research in physical science proposed by the bishop of Ripon, E. A. Burroughs, was without merit. Although the bishop later said that he had made the proposal "with a broad smile," Millikan used the remark as the excuse to expostulate on the evils of inhibiting scientific research. The question the bishop should have asked, Millikan suggested, was whether persons were being educated "to distinguish . . . between the truth of the past and error of the past, and not to pull them both down together?" Millikan was not going to deny that there was a gap between the achievements of scientific research and their public assimilation. It was, however, public education, not science, which had to be reconstructed to close this gap. The public had to learn not to fear science, not to pay attention to men like Soddy, for science disclosed nature's benevolence and was therefore itself benevolent. If humanity was threatened, it was from the neurotic and sexualizing influences of

32. See Herbert L. Sussman, *Victorians and the Machine: The Literary Response to Technology* (Cambridge, Mass.: Harvard University Press, 1968); Millikan, "Alleged Sins of Science," in *Science and the New Civilization*, pp. 58–59, 62–63; Millikan, "The Future of Steel," in ibid., pp. 136–37; Millikan, "Available Energy," in ibid., pp. 95–96.

modern literature, from cubism in art, and from irrational social protest.[33]

Robert Millikan was not the only national scientist to make cosmic progress the basis of scientific progress. Michael I. Pupin, the Columbia University physicist, also brought this thesis before the public in the 1920s. Pupin's scientific philosophy was especially attractive to the American people since he symbolized many American myths—the Serbian immigrant who became a distinguished scientist, whose contributions to the mathematical theory of electrical inductance made possible long-distance electrical transmission and brought a personal fortune as well. Pupin's autobiography, serialized as well as published as a book (1924), promoted the ideology of the national scientists. He believed that the creation of a national science under the guidance of the National Research Council was to usher in the American ideal democracy; this belief was derived from his vision of cosmic progress. Progress, for this scientist, was more than the increasing complexity of man's evolving culture. Progress was "the complete co-ordination of all natural activities, the activities of the atoms in the burning stars as well as of the cells in our terrestial bodies." During its history, science had revealed successively newer realms of coordination in physical reality. The new physics since 1896 had itself revealed the reality of radiant energy in motion.[34]

As the motion of radiant energy indicated, the universe in all its levels of experimental reality was in constant flux: the visibly

33. Edward Arthur Burroughs, bishop of Ripon, to Robert A. Millikan, March 25, 1930, box 39, Millikan Papers. In the paper, "Alleged Sins of Science," in *Science and the New Civilization*, p. 59, Millikan attributed the bishop of Ripon's proposal for a moratorium to a fear of atomic explosives. This was Millikan's misunderstanding, as the bishop pointed out in the letter to Millikan cited here. The ensuing correspondence is in box 39, Millikan Papers. Millikan, "Science and Modern Life," in Francis John McConnell et al., *The Creative Intelligence and Modern Life* (Boulder: University of Colorado Press, 1928), pp. 148–49. An expression of the standard response of scientists to the proposal for a moratorium on research was made by A. W. Meyer, "That Scientific Holiday," *Scientific Monthly* 27 (December 1928): 542–45. Millikan, "Science and Modern Life," in *Science and the New Civilization*, pp. 9–12.

34. Pupin, *From Immigrant to Inventor* (New York: Charles Scribner's Sons, 1923), pp. 30–35, 43–71, 100–06, 330–37, 342, 384–87. Pupin, *The New Reformation: From Physical to Spiritual Realities* (New York: Charles Scribner's Sons, 1927), pp. xiii–xv.

material level, the electrical level, and the radiant level. The special theory of relativity reinforced his conviction of the Heraclitean character of the universe—to each person, the relativity of motion revealed a "primordial flux." The physical objects—shoes to stars—which populate the perceptive world and which appear as eternal permanences were only momentarily stable coordinations, or aggregations, of the materials in the flux. Each realm of aggregations served to coordinate a less broad realm of aggregations. Thus the universe as a whole coordinated the galaxies, the galaxies coordinated solar systems, solar systems coordinated the planets; molecules coordinated atoms, atoms coordinated electrons. Pupin called this coordination "service." The stars thereby served to maintain life by emitting radiant energy.[35] Pupin's cosmology closely resembled Millikan's, for Pupin also believed nature had a bias in favor of man.

The search for cosmic progress discovered a universe fortunately disposed for scientific investigation. The universe was, as Millikan never tired of reiterating, "a mechanism, every part and every movement of which fits in some definite, invariable way into the other parts and the other movements." This classical concept of a mechanical universe did not contradict the concept of a universe in flux. Millikan, like Pupin, would say the universe was a "changing, evolving, dynamic, living organism."[36] Considered as a whole, the universe was a machine in cyclical motion, as depicted in MacMillan's cosmology, with only the parts in change and evolution. The form of the universe remained the same. Radioactivity and the natural transmutation of the elements were physical phenomena analogous to biological evolution, but both, nevertheless, occurred within a steady-state cosmos. The universe as a whole could not be undergoing a true evolution because this would have indicated the validity of the second law of thermodynamics.

True evolution of the cosmos would also imply that the laws of nature changed. If natural laws were descriptions of nature or

35. *The New Reformation*, pp. 176–77, 178, 200.
36. Millikan, "The New Physics," p. 608. "The Birth of Two Ideas," *Scribner's Magazine* 80 (November 1926): 558.

revelations of nature's true conditions, then they would have to evolve with the universe. Evolution of the laws of nature would have meant that the knowledge derived from one cosmic era would not be applicable to another era. This would destroy the scientific basis of progress which depended on the slow accumulation of knowledge which helps each generation more successfully meet its challenges.[37] When Millikan wrote in the 1920s, there was no optical evidence that physical phenomena which occurred hundreds of millions of years earlier obeyed natural laws different from those then valid. This was enough to convince the national scientists of the soundness of their conception of cosmological and scientific progress.

Robert Millikan's Scientific Method

How did Millikan describe the scientific method which revealed the beneficence of nature? He was not forced, as was Slosson, to concentrate on the qualities of the scientist's personality to illustrate the method. Nevertheless, he did believe that certain qualities were important: humility, objectivity, and concern for the common good. Humility was a quality befitting the scientist because of the great discoveries of the new physics. Classical physicists had often been arrogant in their assumption that they knew all there was to be known about nature. The discovery of qualitatively new physical phenomena, such as the X ray and radioactivity, had convinced Millikan that the scientist could never be assured he had a complete knowledge of nature. Humility made scientific objectivity possible. The humble scientist would be willing to set aside a favorite theory in the light of contradictory facts. The dogmatism of traditional doctrines and the tenacity of inherited concepts, such as *mind*, *matter*, *organic*, and *inorganic*, had to be overcome so phenomena could be seen as

37. See Stephen Toulmin and June Goodfield, *The Discovery of Time* (1965; reprint ed., New York: Harper & Row, Torchbooks, 1966), pp. 263–65. The emphasis on the accumulation of valid knowledge was stressed in Millikan, "Science and Social Justice: 'A Stupendous Amount of Woefully Crooked Thinking,' " *Vital Speeches of the Day* 5 (December 1, 1938): 98.

they really were. Millikan also thought that scientific motivation burned longer and brighter when the scientist had concern for the common good of men. For Millikan, this was a peculiarly Christian ideal, but one which was applicable to the scientist. The ideal of the common good could not guide the actual line of scientific research since this direction was dictated by the phenomena. But a fully scientific civilization in which all departments of culture, except religion, were governed by the scientific method could not flourish without this altruistic ideal.[38]

The scientist's method had these main components: the working hypothesis, the probative experiment, the accumulation of facts, and the drawing of deductions. The method was employed most successfully in group research when there was communication between and coordination of the groups. This method can be simply illustrated by references to Millikan's own cosmic ray research. The purpose of the working hypothesis and the probative experiment was to establish the causes of phenomena, that is, the invariant and necessary relationships between several phenomena.[39] The working hypothesis in Millikan's initial balloon experiments (1922) was that the rate of ionization of the electroscope would drop as the balloon instrumentation rose if the suspected rays originated in the earth's surface. The rate of ionization, however, rose. While this was not conclusive evidence that the rays did not come from the surface—since an ad hoc hypothesis might be advanced to bring the observed increase in rate of ionization in line with the working hypothesis—it was enough indication to warrant a more plausible working hypothesis. The new hypothesis, that the rays came from above, was verified by experiments which showed that the rate of ionization decreased with descent in altitude. The final question was whether the rays originated inside or outside the atmosphere. The assumption was

38. Millikan, "I. The Evolution of Twentieth-Century Physics," *Evolution in Science and Religion* (New Haven: Yale University Press, 1927), pp. 9–11, 28; "II. New Truths and Old," in ibid., p. 59. Millikan, "The New Physics," p. 609; "Science and Modern Life," in *The Creative Intelligence*, pp. 158–59; "The Three Great Elements in Human Progress," in *Science and the New Civilization*, the entire essay.

39. Millikan, "The New Physics," p. 609.

made that if the penetrating power of the rays showed a weakening equivalent to a weakening which would occur by their passage through the entire atmosphere, then they could not have originated inside the atmosphere. Experiments made in 1925 verified this last working hypothesis and the rays were designated by Millikan "cosmic" rays.[40]

The probative experiments were used to falsify each working hypothesis about the relationships between rates of ionization, the direction of the rays, and their penetrating power, until the correct hypothesis was reached. The experimentally established relationships became *facts* which were added in Slosson's "bricklayer method" to other such facts. Such facts were never rejected by subsequent scientific work. Only inferences from the facts might later be rejected or modified. "A science," Millikan explained, "grows in the main as does a planet by the process of infinitesimal accretion. . . . Almost every new theory is built like a great medieval cathedral, through the addition by many builders of many different elements."[41]

When a sufficient number of facts had been accumulated, deductions would be drawn. In his cosmic ray research, the major deductions drawn by Millikan were the origination of the rays in interstellar space and their creation as a by-product of atomic synthesis.[42] It was, of course, extremely difficult to draw necessary inferences from the stated facts. It was easier—and this was what Millikan actually did—merely to draw inferences not in conflict with the facts (that is, sufficient). The particular deductions Millikan made were to support his and MacMillan's cosmology. It was impossible, with the few verified facts available about cosmic rays, for Millikan to have made his two deductions logically necessary. To the contrary, it seems obvious that Millikan

40. This was reviewed by Millikan in *Cosmic Rays*, pp. 18–19.
41. Millikan, "New Truths and Old," pp. 50–51; "The Significance of Radium," in *Science and Life* (Boston: Pilgrim Press, 1924), pp. 15–16.
42. Millikan, "Science and Modern Life," in *The Creative Intelligence*, pp. 158–59. Millikan's last hypothesis, in the 1940s, was that cosmic rays originated in interstellar space, but in a process of atomic disintegration, not atomic synthesis. See *Electrons (+ and −), Protons, Photons, Neutrons, Mesotrons, and Cosmic Rays* (Chicago: University of Chicago Press, 1947), ch. 20.

drew the deductions he did, out of the large number of inferences possible, precisely because he wanted to support his particular vision of the cosmos.

Though Millikan's cosmological vision, therefore, was of overarching importance in providing working hypotheses, suggesting lines of research, and designating what kinds of deductions should be drawn from the few facts, the role of such visions and theories in science was not discussed by Millikan, just as it was not discussed in Slosson's popular science. The critical examination of fundamental concepts was no component of Millikan's version of the scientific method, although Einstein's theories certainly made such criticism a basic aspect of modern scientific activity. The omission of theory formation was most peculiar. The reason for the omission, however, is apparent. Millikan and Slosson were presenting a conservative interpretation of science. Their fundamental ideological objective in explaining science to the lay public was to establish the dependence of national progress on science. This required the concept of inevitable progress in science. Millikan, Pupin, and Slosson sought the guarantee for the inevitability of scientific progress in the cosmos. They could not emphasize the importance of the cosmological speculations or the theories precisely because the importance of the speculations made it clear that scientific progress was neither inevitable nor achieved by Baconian techniques. The deductions Millikan drew about the origination of cosmic rays were not necessary and were made to support a cosmology that would provide for human progress.

Conservative, ideological motivation guided the national scientists to a conservative interpretation of science itself. Scientific progress was evolutionary, not revolutionary. Einstein's theories (which, Millikan failed to point out, were not like medieval cathedrals) did not replace Newton's physics; they only modified it. Yet the Einstein theories and the laws of thermodynamics sufficiently threatened the scientists' conservative ideology to force Millikan to leave the drudgery of fact collection and take up cosmological speculation. He did not consider his cosmology as imaginative, but it was. He was in the ironic position of making hasty, qualitative hypotheses in support of a visionary universe in order to defend the view that science was a process of patient accumulation of facts and cautious hypothesizing.

The final aspect of Millikan's scientific method must be discussed for its importance in his social philosophy. Although traditional, American individualism was a basic social value, Millikan was firmly convinced that coordinated group research was necessary for scientific progress. This conviction was born in his First World War experience. The war proved that isolated, individual research, such as Millikan's own prewar determination of the electronic charge, was inadequate to provide national progress and professional scientific progress. The clearest statement of the necessity for group research, still fresh with the impact of the war, was the address "The New Opportunities of Science" (1919). Each belligerent appealed to the native genius of its citizens for scientific ideas and inventions which would help it fight the war. Without exception, these appeals were futile. Neither the amateur scientist, the isolated professional scientist, nor the basement inventor produced anything to aid the war effort. Consequently, Millikan asserted, "it may be set down as a fact fairly well established by the experiences of the Great War that rapid progress in the application of science to any national need is not to be expected in any country which depends, as most countries have done in the past, simply on the *undirected* native genius of its people to make these applications."[43] This was a realization made by the national scientists in the world war and was the motivation for their attempt to institute a national science during the war and in the next decade.

Rapid scientific progress could not be made outside the mission-oriented research groups. In the postwar decade the National Research Council was attempting to nationalize—to coordinate on a national level for a national purpose—the research of America's diverse university scientists, government scientists, and researchers in private and philanthropic institutions. Duplication

43. Millikan, "The New Opportunities in Science," *Science*, n.s. 50 (September 26, 1919): 286.

was to be avoided, research compared, efforts directed to cover all necessary fields.[44]

While the utilization of the scientific method in group research was a lesson of history, it complemented and promoted the other components of the scientific method which were, so to speak, lessons of nature. Group research was admirably suited to the accumulation of facts. But group research was not conducive to theoretical science. Committees and teams did not create visionary cosmologies. Coordinated group research was one aspect of the ideology of national science. This aspect cast also a conservative tone over the ideology, making the ideology inimical to theoretical research. Theory has been, as much from architectonic requirements as from human vanity, the product of the individual mind. Theoretical science became a casualty of the need to guarantee scientific progress. The great theories of modern science, like Einstein's magnificent creations, were a threat not only to the personal investment of the national scientists in classical science, but to the concept of scientific progress, and therefore to the whole ideology of national science.

The Alienation from Nature: Paul R. Heyl

Robert Millikan's search for cosmological progress was undoubtedly a personal success. And this success was shared by Michael Pupin and Edwin Slosson. In the face of apparent threat from the second and third laws of thermodynamics and the relativity theories, these national scientists were able to retain their faith that the cosmos itself sanctioned scientific progress.

The character of this threat to the faith of the scientists can be clarified by turning to the ideas of a scientist who, under the threat of modern physics, suddenly lost that faith. Paul Renno Heyl was a physicist with the Bureau of Standards, whose major scientific work was the precise redetermination of the Newtonian gravitational constant. Heyl was not a scientist of the importance

44. Ibid., pp. 292, 293.

of Hale, Millikan, Pupin, Slosson, and the others who established the national scientific institutions and formulated an ideology of national science in the 1920s. But he was not unknown to the American public. He published frequently in *Scientific Monthly* and *Scientific American*, as well as less popular magazines.

When the Einstein controversy began in 1919, Paul Heyl did not accept the theories of relativity. The theories contravened common sense and had little supporting evidence. But by the early 1920s, the three classic tests of the general theory had forced him to accept the theories, provisionally, as working hypotheses for the investigation of physical nature. He would not follow Eddington and other philosophers of relativity into a theory of nature derived from the general theory. The theories were not final descriptions of physical reality. They only corresponded to experimental facts more closely than Newtonian physics, and Heyl fully expected (and hoped) that they would be replaced by more accurate and philosophically acceptable theories.[45] The theories of relativity were only brilliant, verified summaries and generalizations of experiment.

Heyl's acceptance of the theories of relativity was, therefore, strained and tenuous. He hesitated before the physical reality and the new character of science they seemed to describe. In the mid-1920s, Heyl's adherence to the ideology of science was balanced by his delicate, qualified acceptance of the theories of relativity. Later in the decade, another development in modern science, wave mechanics, shattered this fragile equilibrium and plunged him into despair and rejection of the ideology of science. In its alteration of scientific method and its contradiction of common experience and common sense, wave mechanics had completed the scientific revolution begun by the theories of relativity.

In the mid-1920s, Heyl trusted nature and science. He believed nature was rational, machinelike, and beneficent. He trusted

45. Paul R. Heyl, "Space, Time and Einstein," *Scientific Monthly* 29 (September 1929): 234–35. This was the import of Heyl's statement in "The Common Sense of the Theory of Relativity," ibid., 17 (December 1923): 521–22. Heyl, "The Fundamental Concepts in Physics in the Light of Recent Discoveries," *Science*, n.s. 61 (February 27, 1925): 225; Heyl, *The New Frontiers of Physics* (New York: Appleton, 1930), p. 127.

science to produce eventually an understanding why nature was
as she was and to reduce all phenomena to one reality and one
law. His trust in nature was identical with that of the romantic
poet, and his trust in science was identical with that of the modern philosopher. "We," Heyl wrote of the scientists, "all trust
nature as the solid ground beneath our feet." He approached
nature with the awareness and reverence of a child—the humility
which Millikan said the scientist must have. Heyl was as
entranced by the wonder of a falling stone as by, we must presume, the blowing clover and the falling rain. He was as struck
with awe by the "flickering glow of the vacuum tube" as by the
vault of the night skies. In a literary fantasy which he published,
Heyl described a confrontation between a humanist, who considered science cold and alienating, and himself, the scientist. The
humanist feared the scientific attitude of mind—always logical,
always "schematizing, ordering, classifying, house-cleaning, put-
ting things comfortably to rights." The scientist thought this was
an unfounded fear. In this literary fantasy, Heyl imagined
informing the humanist that this scientific attitude led men, in
the end, to a worship of nature and nature's God. And in Heyl's
imagination, the humanist was stunned by this dramatic revelation and instantly converted to a faith in science. His own literary
talents being inadequate to express his adoration of nature, Heyl
fell back on Longfellow: Nature was an old nurse, singing the
rhymes of the universe to the scientist. [46]

Nature was so constituted that the scientific method could lead
men to a reasonable understanding of her. Despite the bewilder-
ing discoveries of new phenomena since 1895 and the equally
confusing theories of relativity, "we are encouraged to believe

---

46. Heyl, "The Student of Nature," *Scientific Monthly* 24 (June 1927): 498–502; Heyl,
"The Solid Ground of Nature," ibid., 25 (July 1927): 26, 28. These articles appeared only
several months before the announcement by Heisenberg of the uncertainty principle.
Heyl, *The New Frontiers of Physics*, pp. 150–51. Heyl, "The Humanist and I," *Scientific
Monthly* 21 (August 1925): 174–76. In Heyl's "The Solid Ground of Nature," he spoke in
the guise of a philosopher. Heyl, "The Wonder of the Commonplace," *Scientific American*
135 (October 1926): 250. Heyl, "Visions and Dreams of a Scientific Man," *Scientific
Monthly* 26 (June 1928): 515. The Longfellow poem was "Lines on the Fiftieth Birthday
of Agassiz."

that nature as a whole may not be beyond our eventual comprehension. The student of nature is ever optimistic." He did not
think that the physical reality behind the phenomena was inevita-
bly unavailable to human understanding, as the theories of relativity implied. If, as Herbert Spencer had argued in the nineteenth century, this physical reality was unknowable, there would
be no reason to work at science. And Heyl, for one, would leave
the laboratory for work more satisfying to the soul.[47]

The triteness and uncritical character of Heyl's views indicated
the comfortable, inherited faith that he held. His views are easily
traced to the Newtonian world view of the late seventeenth and
eighteenth centuries and to a superficial reading of the romantic
poets of the early nineteenth century. For Heyl the scientific
method had the objective of revelation. This objective emphasized one aspect of the dual character which the scientific method
had had since the eighteenth century. The two aspects of this
dual character were manipulation and revelation. Manipulation
had been a part of the scientific method since the English publi-
cist of science, Francis Bacon, for whom the purpose of this method was to provide material progress by the manipulation of the
forces and phenomena of nature. Revelation became a part of the
scientific method with the Protestant Reformation and especially
with the Newtonian world view. The method of science provided
the Protestants with revelation of God's will in nature at the time
they were denied the Catholic tradition of revelation and the
sacraments. With the triumph of the Newtonian world view, the
character of the scientific method as revelation became formally
elaborated into the natural theology. The revelatory aspect of the
scientific method was very strongly emphasized in America in the
nineteenth century as a main justification for science to the public. In the later nineteenth century, the manipulatory character of
the scientific method became more important as the nation
industrialized and as scientific positivism (especially in France
and Germany) undercut the scientific power of revelation. The

47. Heyl, "The Student of Nature," pp. 500–01. Heyl, *The New Frontiers of Physics*,
pp. 150–51.

American physicists, who represented one of the last elements in the scientific community to feel the impact of the industrial revolution (much later than, say, the chemists and geologists), retained the strong faith in the revelatory character of science into the present century. For Heyl, in the mid-1920s, the purpose of science remained primarily and literally to open men's eyes, to reveal, rather than to manipulate.[48]

The ultimate threat of the theories of relativity was to deny the scientific method the power of revelation; in other words, to make the scientific method merely manipulative. Science became a technology, mediating between men's minds and a physical reality which could not be revealed to them. This technocratization of science has been one of the great themes of twentieth-century intellectual history, as the French intellectual historian, Jacques Ellul, has so effectively argued.[49]

This threat created by the theories of relativity, which Heyl had held away by considering the theories as only working hypotheses, was finally consummated by the development of wave mechanics in the late 1920s. Heyl's fragile ideology of science broke apart. Erwin Schrödinger had conceptualized the atom as a vibrating wave which was not substantial. Werner Heisenberg proved that the position and the momentum of a moving particle could not simultaneously be known. On the atomic level, the universe was indeterminate. Finally, Paul Dirac proposed that the fundamental things of nature were inexpressible by numbers. "Numerical relations begin to appear only when

48. On the revelatory character of the scientific method, see Carl L. Becker, *The Heavenly City of the Eighteenth-Century Philosophers* (1932; reprint ed., New Haven: Yale University Press, 1964), ch. 2; Ernst Cassirer, *The Philosophy of the Enlightenment*, trans. Fritz C. A. Koelln and James P. Pettegrove (1932; reprint ed., Boston: Beacon Press, 1955), ch. 2. The importance of the revelatory character of the scientific method in America is illustrated in the career of Louis Agassiz and in the justification of the scientific method to the public. See Edward Lurie, *Louis Agassiz: A Life in Science* (1960; abridged ed., Chicago: University of Chicago Press, Phoenix Books, 1966), and George H. Daniels, *American Science in the Age of Jackson* (New York: Columbia University Press, 1968).

49. Jacques Ellul, *The Technological Society*, trans. John Wilkinson (1954; reprint ed., New York: Random House, Vintage Books, 1967), pp. 42–60, on the technocratization of science owing to the industrial revolution. On the change in the concept of method, see the important work by Wylie Sypher, *Literature and Technology: The Alien Vision* (New York: Random House, 1968).

we reach combinations of these fundamentals of a certain degree of complexity." Dirac had destroyed the "solid ground of nature." Nature could not be known entirely by science. This, in 1931, was too much for Heyl. Perhaps, he wrote, the bishop of Ripon was right: Science did need a holiday. The new discoveries and theories had been made so rapidly that they could not be assimilated. The very ability of science to know nature was now being denied. Science had become a " 'cloud-cuckoo land.' " Nature was unreal.[50]

Heyl felt obliged, if in despair, to accept the new scientific views. They did, after all, meet the new requirements of physical theory which the theories of relativity had established. If they did not provide an understanding of nature behind the phenomena, or if, indeed, they indicated that nature was not rationally understandable, they did provide practical experiments and more exact mathematical explanation of the experiments. This was the direction of successful physics, and Heyl could not withstand it. But the price for the change was high: The rational universe of natural law and order was gone.[51]

Heyl drew the implications following from acceptance of the new physics in his book *The Philosophy of a Scientific Man* (1933). The chapters of this book were rewritten versions of his earlier papers on the scientific method and the difference in wording between the earlier essays and the chapters defined precisely Heyl's intellectual crisis.[52] In *The Philosophy of a Scientif-*

50. These developments have been reviewed in Werner Heisenberg, *Physics and Philosophy: The Revolution in Modern Science*, Gifford Lectures, University of St. Andrews, 1955–56 (New York: Harper & Row, Torchbooks, 1958). Also relevant is Niels Bohr, *Atomic Theory and the Description of Nature* (Cambridge, England: University Press, 1961), "The Atomic Theory and the Fundamental Principles Underlying the Description of Nature." Both of these architects of modern quantum theory ascribe to the theories of relativity the basic scientific revolution concerning the relationship of science to reality. The quantum theory, therefore, appears only as a consummation of this aspect of the scientific revolution, and is the reason why I have not felt it necessary to discuss the theory in detail. Heyl, "The Perspective of Modern Physics," *Scientific American* 145 (September 1931): 168–70.

51. Heyl, "The Perspective of Modern Physics," pp. 168–70. Heyl, "Cause or Chance?" *Scientific Monthly* 34 (March 1932): 273–74; Heyl, "Romance or Science?" *Journal of the Washington Academy of Sciences* 23 (February 1933): 82.

52. The different versions of Heyl's papers were these: "The Mystery of Evil," *Open Court* 34 (January–March 1920): 34–48, 74–86, 155–69; ch. 3, "The Dual Aspect of

ic Man, Heyl asked two questions forced on him by the new physics. What is the place of reason in nature; that is, does reason represent "even qualitatively the principles to which the actions of the outside universe correspond"? Second, on what basis can we hope reason might ever so represent the principles of the outside universe? Heyl, like Slosson and Millikan in the 1920s, struggled to maintain the integrity of reason against the attacks of irrationalism. Despite the difficulties it encountered, reason was the highest product of man's evolution. It remained man's safest guide through the universe.[53]

Not alone did the Freudian concepts of the unconscious mind and the libido or Watsonian behaviorism threaten the adequacy of reason. Physics itself carried this threat. Modern physics implied that the rationality which scientific reason had once discovered in nature was no more than the projection of itself into nature. Science, which had been considered above religion and superstition, was in this respect exactly on their level. The superstitious Australian aborigine, the medieval theologian, and the modern scientist were all alike in projecting their reason into nature. Only "the forms in which they shape the mystery may differ." Science had rejected the anthropomorphism of religion; now the theories of relativity and wave mechanics exposed and rejected the subtle anthropomorphism in science itself. The great lesson of the theories of relativity was that reason cannot have access to the reality behind the phenomena. Any understanding which the mind thinks it has of reality was only the vain reflection of itself in nature.[54]

Mind had no logical justification for projecting itself into nature. It did so only out of an emotional need to organize what it perceived and to remove the fear of chaos. This emotional basis

---

Nature," and ch. 4, "The First Alternative: The Cosmic Soul," in *The Philosophy of a Scientific Man* (New York: Vanguard Press, 1933). "The Student of Nature" (1927); ch. 5, "The Second Alternative," in *The Philosophy of a Scientific Man.* "The Solid Ground of Nature" (1927); ch. 2, "The Dual Aspect of Nature," in *The Philosophy of a Scientific Man.* See also: "Common Sense of the Theory of Relativity" (1923); ch. 2, *The New Frontiers of Physics* (1930).

53. Heyl, *The Philosophy of a Scientific Man*, pp. 20–21, 31–33, 164.
54. Ibid., pp. 34–35, 39, 154–55, 166–67.

for science was no more, no less, valid than the emotional basis of superstition and religion. Science had never really perceived the handiwork of the Deity in nature. It had perceived only the handiwork of itself.[55]

The answers to Heyl's queries were clear. Reason had no place in nature. Nature was not reasonable or machinelike and possessed none of the qualities of reason. Nature was neither good nor evil. Nature was not conducive to progress. And nature might very well be capricious. Nature and science had been deprived of all anthropomorphic qualities. Science had gone, said Heyl, using Nietzsche's phrase, beyond good and evil.[56] It was nonsensical to say, as had Millikan and the national ideologists of science, that nature was good, guaranteed progress, and could not be dangerous, or even that the scientific understanding progressed.

Robert Millikan's search for the cosmic guarantee of scientific progress as the basis of the ideology of national science ended in Paul Heyl's courageous admission that the cosmos was indifferent to the progress of science and to the progress of men. Heyl himself had only a hope that some day an understanding of nature—a sudden revelation of the unity of the universe—would be achieved. But this hope was not, he said, based on science. It was a faith that transcended science.[57]

55. Ibid., ch. 5.
56. Ibid., pp. 154–55, 166.
57. Ibid., p. 168.

(6) IDEOLOGY: SOCIAL VALUES

The national scientists emerged from the First World War into a confusing social situation. The war experience brought a realization of the advantages of national science. They thought this new organization could be supported in the peace only if the scientific and nonscientific sectors of American culture were united. But the prewar progressive liberalism which had placed great social and political value on science was in disillusion and disarray after the war. The scientists sought to reunify American culture by overcoming the alienation of the nonscientific public from science through popularization of the method and values of science. They sought to guarantee the progress of civilization in the scientific method and to guarantee the progress of science itself in the character of the cosmos. But this was not enough. If culture was to be unified, the ideology of national science had to guarantee the preservation of traditional social values.

Science: The Salvation of Democracy

The national scientists expressed confusion immediately after the war about the relationship of science to the other sectors of American society. All found their way out of this confusion in the 1920s in the belief that science best promoted traditional democratic values. George Ellery Hale, speaking in a symposium on "Scientific Education in a Democracy," admitted that no one seemed certain of the scientist's responsibilities on the return to peace. Should scientists return to the disinterested pursuit of natural knowledge or enter industry? What value should they and society place on science? To answer that science was practical did not guide the scientist to his proper role. Hale thought that the distinction between pure science and its applications should be broken down: "Only thus can the highest advantage of science and industry, the chief interests of public welfare and the greatest national progress be attained."[1]

Robert Millikan shared some of Hale's uncertainty concerning the exact social role of science. In an address before the summer session of the University of Chicago in 1919, Millikan discussed the opportunities for science created by the cooperative research projects of the war. The world had been awakened to the potential of research. Science had a role in industrial development, in the creation of wealth, and in molding personal character. But how this role was to be played, he did not know. The mobilization of scientists in the war had given them a sense of high social worth, but there were no projects for scientists in the peace.[2]

A thoughtful and articulate expression of the uncertainty concerning the exact social role of science was made by John Campbell Merriam, a paleontologist and president of the Carnegie Institution, as well as a friend of Hale and Millikan. The peace required the reconstruction of society, he said, but it set no schedule for doing so. Nevertheless, it was urgent to understand what questions were to be asked about science and society and to know who was responsible for their solution. Science would be deeply implicated in this reconstruction because all constructive action required prior research.[3]

The scientists agreed, in general terms, that the role of science in America was to maintain democracy. The immediate origin of this agreement was their experience of the war in which science

1. George Ellery Hale, "The Responsibilities of the Scientist," *Science*, n.s. 50 (August 15, 1919): 143, 146.

2. Robert A. Millikan, "The New Opportunities in Science," ibid. (September 26, 1919): 285–87.

3. John Campbell Merriam, "The Research Spirit in Everyday Life of the Average Man" (Address delivered as retiring president of the Pacific Division, American Association for the Advancement of Science, Seattle, June 17, 1920) *Published Papers and Addresses* (Washington: The Carnegie Institution of Washington, 1938), 4: 2377, 2380. *(Published Papers and/Addresses* cited hereafter as *PPA.)*

had made its contribution to preserving democracy. The less immediate origin of their agreement was the value given to science in progressivism. Millikan often stated in his public addresses that American democracy would work only if everyone took a more rational, objective, and scientific approach to living. Radical reform of democracy would not be necessary if the contributions of the physicists, chemists, biologists, and engineers were not resisted. The scientific method "represents the only hope of the race of ultimately getting out of the jungle."[4] John Merriam also thought democratic government was dependent on science. Democracy had to be government by judgment rather than by arbitrary decision. Scientific method and research were necessary to guarantee the validity of judgment. The public had, therefore, to be educated in science.[5] Michael I. Pupin believed that science was to raise America to the ideal democracy. He fought the criticisms made, for example, by André Siegfried in *America Comes of Age* (English edition, 1927) that science had fostered materialism and had debased America.[6] Democracy itself could be dangerous, as the world war demonstrated, unless scientific reason was its guide. Only science, Pupin maintained, could "make the world safe for democracy."[7] Edwin Slosson thought scientific progress was the basis for the progress of material democracy. And William Emerson Ritter thought science should halt the economic and spiritual degeneration of American democracy.[8]

Statements that only science or the scientific method could

4. Millikan, "Science and Human Affairs—Abstract," *Addresses and Proceedings, National Education Association* 61 (1923): 847. Millikan, "Science and Society," in *Science and Life* (Boston: Pilgrim Press, 1924), p. 81. Millikan, "Science and Modern Life," in Francis McConnell et al., *The Creative Intelligence and Modern Life* (Boulder: University of Colorado Press, 1928), p. 159.
5. Merriam, "Making a Living—or Living?" (Address delivered at the 98th commencement of New York University, June 1930), *PPA* 4: 2036.
6. Michael I. Pupin, *From Immigrant to Inventor* (New York: Charles Scribner's Sons, 1923), pp. 321, 341–42; Pupin, *Romance of the Machine* (New York: Charles Scribner's Sons, 1930), pp. 7–26, 94.
7. Pupin, *From Immigrant to Inventor*, p. 273.
8. William Emerson Ritter to Vernon Lyman Kellogg, December 7, 1921, box 153, John Campbell Merriam Papers, Manuscript Division, Library of Congress (cited hereafter as Merriam Papers); see also note 25, ch. 3.

save American democracy would be, however, both hollow and a cheap route out of social confusion if they were not accompanied by at least a rudimentary analysis which demonstrated science's relationship to social unity. This analysis was most explicitly set out in the 1920s by Robert Millikan and John Merriam.

Robert Millikan explained on several occasions that the unity of the values of science and the values of democracy derived from the assumption that the scientific method was also a social method.[9] What could Millikan have meant by this assumption, since he does not develop it deeply? There are several possibilities. The assumption could mean that the technique of scientific method was social in character, finding its genesis and terminus in public situations. Or, the assumption could mean that the scientific method, regardless of whether its technique was social, was suited to the solution of problems arising in a particular kind of society. The latter possibility raises the question whether the suitability of the scientific method for solving social problems derived from the character of American society alone, just as Millikan thought the suitability of the scientific method for the investigation of physical nature derived from the special character of nature itself.

Edwin Slosson and Robert Millikan both interpreted the scientific method as social because its origin and conclusion were in social situations. Furthermore, the greatest scientific progress was achieved when scientists worked in research groups. Slosson's and Millikan's interpretation was derived from John Dewey's philosophy and less directly from the pragmatic tradition generally. For Dewey, the scientific method was a means of releasing psychological tension generated by discordance in a person's social situation. For Charles S. Peirce, one of the founders of pragmatism, the scientific method was social in a more profound way. The scientific method was a means by which men reached agreed-on truths which were independent of what any one scientist thought about them. Scientific language was social because its purpose

---

9. Millikan, "The New Physics," *School Review* 23 (November 1915): 609; "Science and Society," in *Science and Life*, p. 71.

was to symbolize the conceivable effects of any event. For Peirce these conceivable effects were not private and psychological as they tended to be in William James's philosophy, but could be witnessed by anyone. Experimentation was a method of reaching agreement on what were an event's conceivable effects by probing and manipulating nature. For Peirce, then, the scientific method was a social method because it placed the meanings of statements in a realm where all men could agree on them.[10] In the 1930s, Robert Millikan was to describe a "jury method" for solving social problems that was analogous to Peirce's method of science.

For Millikan, the scientific method was social also because it worked more fruitfully in mission-oriented group research than in the research of the isolated investigator. The group research in antisubmarine warfare in the First World War had provided greater progress than the occasional contributions of the isolated scientist. This was not an essential modification of the scientific method as a technique. It was more a lesson of efficiency from the scientifically managed factory.

The notion that the scientific method was social because its technique was social in character was not, however, the more important of the two possible meanings of Millikan's assumption that the scientific method was a social method. Rather, the more important meaning was that American democracy was peculiarly suited to the scientific resolution of its social issues. The importance of this latter meaning derived from the inability to conduct scientific experiments, as one component of the scientific method, directly on society. The experiment, as a manipulation of social phenomena for the purpose of fixing meanings and determining agreement, was not available to social scientists. People have been manipulated throughout history, but the scientific purpose for doing so has seldom been acceptable (outside of medical

---

10. Peirce's views on these matters were adequately explained in his famous essays, "How to Make Our Ideas Clear" and "What Pragmatism Is." These essays have been reprinted in *Values in a Universe of Chance: Selected Writings of Charles S. Peirce*, ed. Philip P. Weiner (Garden City, N.Y.: Doubleday, Anchor Books, 1958), pp. 113–36, 180–202.

research). Scientists studying society have had to develop statistical techniques to compensate for this inability to do direct experimentation. But statistics alone, of course, have not always been successful in bringing social scientists to agreement. Millikan meant that the democratic character of American society compensated for the inability to conduct scientific, social experimentation. Millikan was reaching back to prewar progressivism for the assurance that though scientific experimentation could be no part of the scientific method in solving social problems, democracy itself made such experimentation unnecessary. The scientific method as social method relied not on experimentation but on the integrity of judgment of the scientifically trained, individual citizen. "The scientific spirit," as Walter Lippmann said in *Drift and Mastery*, was "the discipline of democracy."[11] Science flourished best in a society which valued individualism, free and private initiative, and mass education, and in turn these were the values most emphatically promoted by science.

The scientific method, therefore, became the basis for the unity of the two cultures. No national scientist proclaimed this theme with greater clarity than John Merriam. For Merriam, as for Millikan, the First World War had proved that scientific "research" (by which he meant the scientific method) was not distinct from the national interest or from the general human interest. Science could be organized to provide security in war and economic progress in peace. Writing of prewar Germany, Merriam mentioned "the strength of German military organization, backed by scientific and economic interests *welded into one powerful instrument.*" This "one powerful instrument" was, of course, what the scientists hoped to create in peacetime America. Unity of scientific and nonscientific cultures was demanded in peace no less than in war. Recognizing that prewar America was gone forever, its division into separate cultures impossible if progress was to be achieved, Merriam called for a "reconstruction" of American society based on the scientific method. Only this method could

11. Lippmann, *Drift and Mastery* (Englewood Cliffs, N.J.: Prentice-Hall, 1961), p. 151.

correctly answer questions, justly resolve issues for all interests, distribute authority, and provide cultural unity.[12]

Research was not simply "the detailed investigations of fundamental scientific principles," but comprised "all inquiry which may be included within the range of thought leading to constructive action." There were two categories of research: research for the discovery of fundamental principles (including nonscientific principles) and research into the methods by which principles were applied practically.[13] This characterization of research certainly did not emphasize experimentation. Rather, it emphasized observational fact-gathering which was applicable to social problems.

Merriam believed research would discover laws of culture analogous to the natural laws discovered by physical science—though he apparently did not consider the logical status of such cultural laws. The expectation that cultural laws could be discovered, providing continuity in change and unity in diversity, was ultimately derived from the eighteenth-century Enlightenment and the triumph of uniformitarianism. The basis of the social sciences in pre-Darwinian intellectual Europe had been in the belief that the method of physics could be applied to society and history. Serious methodological problems (such as the inability to do experimentation) and category mistakes had prevented such a science of society, but these were remedied with the absorption of Darwinian theory into the social sciences and the rise of modern liberalism. Merriam's faith that cultural laws were discoverable stemmed immediately from the general impact of evolutionary theory. The absorption of evolutionary theory—or "reform Darwinism"—into American liberalism helps to account for Merriam's and especially Millikan's belief that American society was particularly suited to the scientific method.[14]

---

12. Merriam, "The Research Spirit," pp. 2376–77. Italics added.
13. Ibid., pp. 2377, 2381.
14. Merriam, "Common Aims of Culture and Research in the University" (1922), *PPA* 4: 2389–90; "Medicine and the Evolution of Society" (1926), *PPA* 4: 2397. See F. A. Hayek, *The Counter-Revolution of Science: Studies on the Abuse of Reason* (1955; reprint ed., London: Collier-Macmillan, 1964).

The research into applications found its most common expression in the daily life of every person and its highest expression, Merriam asserted, in the "great" engineering laboratories of private industry.[15] Both kinds of this research were simply examples of induction based on empirical evidence and inference from the principle to the case. Merriam was trying to exclude trial-and-error invention as a technique of applied research. Why did he choose to describe such traditional modes of reasoning as induction and deduction as "research"? Apparently Merriam was attempting to attach normative significance to such reasoning. Research, it if were to be a basis of cultural unity, had to be a concept broad enough to connote nonscientific as well as scientific judgments (using *scientific* in the professional sense). This was not an unimportant intellectual problem. Science had often prompted an attitude of skepticism toward traditional values and beliefs. It was not clear to all Merriam's contemporaries that ethical values had other than an unfortunate relationship with scientific values. The controversy over evolution in the nineteenth century had been long and traditional values were severely strained before their final synthesis with evolutionary theory late in the century. Only the fundamentalist controversy of the 1920s need be recalled to realize that large segments of the population still refused to accept this synthesis.

Research was a unifying principle, then, because it involved "a reaching out to bring together, organize and interpret whatever may be added to our store of knowledge." The cultured state of mind and the scientific state of mind became alike a "comprehensive vision." Merriam discussed this comprehensive vision explicitly with regard to education. Scientific studies should lead to an understanding of the interrelatedness of all phenomena through law and the continuity of law through time. Similarly, humanistic studies had to establish the "continuities of interests and responsibilities in the world of human life." These interests and responsibilities were shared by science and other sectors of society because all true research was constructive, yield-

15. Merriam, "The Research Spirit," pp. 2381–82.

ing service to society.[16] This was the same concept of service which Michael Pupin made the basis of his concept of creative coordination of democracy by science. As Millikan stated more bluntly, the method of science was the salvation of democracy.

Whether science was the salvation of democracy depended, however, on more than the utilization of the scientific method for the solution of social problems. The research method had specifically to guarantee the existence of the traditional social values of individualism, the private creation of wealth, the democratic distribution of wealth, and education.[17] Only this guarantee could unify the two cultures.

Science and the Preservation of Individualism

Herbert Hoover was a true hero to the national scientists in the 1920s. Many of them were his personal friends. He became the symbol of the reasonableness and realism of the social philosophy they held. Robert Millikan joined Hoover's presidential campaign in 1928 to make a speech in which he extolled Hoover as "one of the most liberal, progressive, wise, and far-seeing men whom we have in American life today, incomparably superior in training, in outlook, in ideals, in constructive power, in background [to Smith]."[18] This sentiment, when the rhetorical overstatement was removed, was shared by other national scientists.[19]

Hoover's *American Individualism* (1922) was the paradigmatic expression of the national scientists' social views. The social philosophy presented in this little, bare work represented a distillation of prewar Republican center progressivism. It was a frank reassertion of this progressivism in reaction to the rise of revolu-

16. Merriam, "Common Aims of Culture and Research," pp. 2387–89. Merriam, "The Research Spirit," p. 2380.

17. Millikan specified these values as the ones to which science could make guarantees in "The New Opportunities in Science," pp. 293–94. For Pupin and Slosson, these were central values. Later in this chapter, Merriam's concern with them shall be discussed.

18. Millikan, untitled ms., dated October 31, 1928, box 48, Robert A. Millikan Papers, California Institute of Technology Archives (cited hereafter as Millikan Papers).

19. See Merriam, "Science and Government" (Address before the Interstate Legislative Assembly, Washington, D.C., February 4, 1931), *PPA* 4: 2477; Vernon Kellogg to Merriam, March 20, 1928, boxes 104, 105, Merriam Papers.

tionary socialism in eastern Europe and bolshevism in Russia. As a social philosophy on the defensive, it was appropriately limited to supposedly indisputable essentials, thereby revealing the apparently conservative position into which science was placed with this philosophy.

Hoover's objectives were to expose the fundamental character of American society and to indicate how this character could be preserved. "Now," he wrote, striking the dominant tone of the book, "as the storm of war, of revolution and of emotion subsides there is left even with us of the United States much unrest, much discontent with the surer forces of human advancement."[20] The national unity of the war had encouraged in many men the hope that this unity might continue after the war, directed in the peace against the ills of national welfare. This was the hope of the national scientists. They sought to preserve this unity by creating a national consensus on science. Hoover's objective was similar: to remind Americans that their commitment to traditional individualism was the best means of fulfilling their hope. Hoover's and the scientists' objectives were entirely complementary.

Hoover's creed of American individualism was simple enough to stand in a self-explanatory quotation:

> While we build our society upon the attainment of the individual, we shall safeguard to every individual an equality of opportunity to take that position in the community to which his intelligence, character, ability, and ambition entitle him; that we keep the social solution free from frozen strata of classes; that we shall stimulate effort of each individual to achievement; that through an enlarging sense of responsibility and understanding we shall assist him to this achievement; while he in turn must stand up to the emery wheel of competition.[21]

This creed depended, as had the progressivism of the professional middle class before the war, on the intellectual qualities of

20. Herbert Hoover, *American Individualism* (Garden City, N.Y.: Doubleday, Page, 1922), pp. 2–3. See also p. 30.
21. Ibid., pp. 9–10.

the individual which were his protection against the economic power of huge industries and unions or the political power of the interests in government. The "one source of human progress" was the opportunity of each person to develop fully his abilities. The individual was protected against the forces seeking to dominate him, to freeze the social environment, not only by his native wits, but by a tradition of idealistic service—"service to those with whom we come in contact, service to the nation, and service to the world itself."[22] This altruism countered the centrifugal tendencies of extreme individualism by promoting cooperation to meet adversity and to aid in individual effort. Pupin and Merriam meant by service what Hoover meant. Robert Millikan undoubtedly intended the same when he asserted that science should be guided by concern for the common good. Science, according to Millikan, determined what was the common good and also demanded the subordination of the individual to the group goal "as a duty—for the sake of world progress."[23] The ethical concern for the common good was similar to the prewar hope of John Dewey that scientific intellectuals could rise above the competition of special interests in society to guide the people to their common benefit.[24] While Hoover's anxiety that individualism should be tempered by social idealism sprang from a fear that in peacetime American individualism might dissolve into anarchy, especially under pressure from competing alien social philosophies, the worry of the national scientists was that the war-born unification of the scientific community, and the community with the rest of society, would disintegrate in the peace.

For Hoover and the national scientists, democracy was only the political expression of social and economic individualism. Democracy was not an equality of condition, but an openness of opportunity. Democratic government reinforced the service ideal by

22. Ibid., pp. 13, 29.

23. Millikan, *Evolution in Science and Religion* (New Haven: Yale University Press, 1927), p. 83. Similar sentiments were voiced also in *Science and the New Civilization* (New York: Charles Scribner's Sons, 1930), pp. 169–70.

24. Dewey used the term *common good* in his *Ethics* (1908). See Sidney Kaplan, "Social Engineers as Saviors: Effects of World War I on Some American Liberals," *Journal of the History of Ideas* 17 (June 1956): 349.

curbing oppressive economic forces, aiding the individual with education, and insuring a fair distribution of products.[25]

What connection did Hoover see between this tradition of American individualism and science? The maintenance of this individualism and democracy necessitated continual creation of new opportunities for the advancement of the individual. As particular social or economic paths became blocked to personal advancement because of the end of the frontier or the saturation of a market, for example, new frontiers and markets had to be created. Hoover could not admit with the pessimists that such new opportunities would eventually cease to be generated. Science, Hoover said, could create endless opportunities. "The days of the pioneer are not over. There are continents of human welfare of which we have penetrated only the coastal plain. The great continent of science is as yet explored only on its borders, and it is only the pioneer who will penetrate the frontier in the quest of new worlds to conquer."[26] A new faith had entered the lexicon of American myths: science, the great unexplored continent, offered by Hoover in 1922. "Science—The Nation's Inexhaustible Reserve," went the title of a Millikan speech in 1930.[27] "Science, The Endless Frontier," wrote Vannevar Bush in 1945. While the prewar progressive faith in democracy and individualism had found its strength in many sources, following the war that faith was drawn basically from science.

The national scientists did not leave the dependency of individualism upon scientific research with Hoover. They sought to demonstrate that science favored an individualistic social system more than any other social system. Science did this because science developed the moral capacity and intellectual capacity of men, solved the problems of distribution of wealth without appeal to socialism, promoted the free and private creation of wealth, and advanced education.

Edwin Slosson had promulgated an image of the scientist as a

25. Hoover, *American Individualism*, pp. 10–11.
26. Ibid., pp. 63–64.
27. Millikan, "Science—The Nation's Inexhaustible Reserve" (Address delivered at the Twenty-Fourth Annual Convention of the Association of Life Insurance Presidents in New York City, December 12, 1930), box 48, Millikan Papers.

particularly virtuous personality. For the national scientists, it followed from this image that to the extent a man was scientific, he was good. The world war, the social chaos and revolution following the war, and the decline of church attendance and moralism during the 1920s contributed to the impression that morality no longer had a substantial basis. The scientists sought to provide that basis by anchoring certain moral virtues to the scientific method itself.

Millikan and Merriam both explained the scientific basis of moral amelioration in addresses delivered shortly after the war. According to Millikan, the fundamental virtues sprang out of the manner in which a scientist approached nature to obtain knowledge. The successful use of the scientific method required true humility; the scientist who was arrogant would miss the discovery of new phenomena which might contradict his favorite theory. Scientific training taught emotional and individual discipline. "From my point of view," Millikan told an audience of teachers, "there is no training in objective, analytical thinking, nor in honesty and soundness of judgment, which is comparable to the training furnished by the physical sciences." Millikan could sound like the pragmatists: "Life presents to each of us one continuous succession of problems to be solved." And science was the best training in problem-solving. In an address of tribute to America's great inventor, Thomas A. Edison, Millikan enumerated the virtues which arose from being a scientist, "namely, modesty, simplicity, straightforwardness, objectiveness, industry, honesty, human sympathy, altruism, reverence and a keen sense of social responsibility." These virtues flowed from the requirements of making a correct scientific judgment. Arrogance, complexity of character, deviousness, subjectivity, laziness, deceitfulness, human callousness, immoralism, and social irresponsibility would all prevent the scientist from gaining "a correct understanding of relations between phenomena, social as well as physical, including that of one's own position in the scheme of things."[28]

28. Millikan, "Science and Human Affairs," pp. 846–47. Millikan, "Edison as a Scientist," *Science*, n.s. 75 (January 15, 1932): 68.

The qualities of character and mind which came out of the scientific method were, of course, precisely those qualities necessary for an individualistic, middle-class society. Few qualities were better suited to preparing the spirit for a contest against the emery wheel of competition. John Merriam, as much as Millikan, was aware of this. "There is no doubt," Merriam told America's average man, "that properly organized and coordinated efforts of science and education may increase greatly the present opportunity of the average man for constructive activity, making his life more useful and happier."[29] Throughout Merriam's public addresses, this theme repeated itself. An education—by which he meant scientific training—would help the individual define his capacities and abilities, provide him with accumulated knowledge prerequisite to using his abilities, train his judgment, enlarge his creativity or his ability to manipulate situations in new ways, and provide a firm basis for moral character.[30] Each person understood his own capacities and, therefore, could move into that social and economic position for which he was best suited. The government need not direct or regulate personal activities. Each person was educated and, therefore, could select the best-qualified leaders—this made representative democracy possible. And the education increased the opportunity of each person to live in a personally satisfying style.[31] Merriam did not think education in scientific research was useful only to the middle-class man on the make. A higher value of this education was in deepening the spiritual worth of an individual and stimulating simultaneously a reaching-out from one person to another in service. Scientific education helped everyone perceive the greatest common good for American society. He would realize that individualism tempered by cooperation would best secure this.[32] The experience of

29. Merriam, "The Research Spirit," p. 2384.
30. For example, ibid., pp. 2379–80, and Merriam, "The Breadth of an Education" (Founders' Day Address at the University of Virginia, April 13, 1922), PPA 4: 2063.
31. Merriam, "Making a Living—or Living?" p. 2037.
32. For example, Merriam, "Spiritual Values and the Constructive Life" (Address at a conference of universities under the auspices of New York University, New York, November 15–17, 1932), PPA 4: 2052–54; Merriam, "Science and Government," p. 2477.

the national scientists in the war was not lost on Merriam, and he was anxious to pass on its lessons to fellow citizens.

It had been a common criticism, especially during the struggle over evolutionary theory, that science promoted philosophical materialism which in turn dried up the human spirit. Certainly, this criticism was true of the nineteenth-century tendency toward physicalism in, for example, the biological work of Jacques Loeb. Though the national scientists like Millikan were reluctant to accept the scientific revolution wrought by the theories of relativity, they did recognize that this revolution had demolished nineteenth-century scientific materialism. Relativity and quantum theories had described the existence of irreducible, qualitatively distinct levels of physical reality. The theories thereby thwarted the effort of materialists to reduce all phenomena, including human experience and consciousness, to the motions of atoms. Without necessarily endorsing the other philosophical implications of the new physics, the national scientists would point out that materialism no longer threatened man's spiritual life.[33]

Science and the Creation of Wealth

Herbert Hoover's new frontier of scientific research, which was to perpetuate individualism, was not limited to this value alone. One major effort of the national scientists was to demonstrate that scientific research made possible the free and private creation of wealth and thereby kept open the economic avenues of individualism. In an address to the New York Chamber of Commerce in 1928, Millikan asserted baldly, "Pure science begat modern industry." Though this was not historically true for a majority of industries, he drew forth the still familiar examples of the electrical and radio industries which developed after Faraday's and Hertz's discoveries. Free basic research into the phe-

---

33. See Millikan, "The New Physics," p. 609; *Science and the New Civilization*, pp. 167, 185–86; *Science and Life*; and ch. 3, *Time, Matter, and Values* (Chapel Hill: University of North Carolina Press, 1932), for a general discussion of the compatibility of scientific and religious values.

nomena of nature must yield discoveries some of which would have economic value. At the same time, such research would indicate what phenomena would have no practical value. This designation of the impractical was almost as important as the discovery of the practical because it prevented industries from wasting capital on useless ventures and could give them a basic surety of the value of their own product. Thus Millikan assured the manufacturers of steel that no other basic metal would replace steel in structure, transportation, or power. This assurance was based on a knowledge of the molecular properties of the elements and could not be shaken. Millikan could lecture to the steel manufacturers if he had to. All of the advances in steel technology, he told them, were dependent on pure research, such as that into the properties of alloys. This research was conducted within the framework of a free and individualistic society. Industrial wealth had been created as much by individual scientific initiative as by individual economic initiative.[34]

There was every promise that this beneficial relationship would last. The application of science to production was increasing automation, freeing persons for service, including science and education. This should create new wealth and new wants.[35]

There was a reason, indebted to prewar progressivism, why individualism was guaranteed by the private creation of wealth. This was the continual increase of the total amount of wealth due to science. The continual increase removed the necessity for a socialist redistribution of wealth. "The method of science," Millikan said reassuringly, "is . . . to create more income and thus [to] have more to distribute." Although this remark was made in 1938 in opposition to the New Deal, similar sentiments had been expressed following the First World War in response to domestic turmoil and European revolutions. "It is probable," Millikan told a University of Chicago audience in 1919, "that the total possibilities of improvement of conditions through distribution are

34. Millikan, *Science and the New Civilization*, pp. 39, 139–41, 147.

35. Millikan, "Education and Unemployment," *Atlantic Monthly* 148 (December 1931): 806.

very limited, while possibilities of improvement through increases in production are incalculable." Europe may have been in social and economic chaos following the war because of the degraded position of the worker, but America could avoid this upheaval because the American worker was more productive than his European counterpart.[36]

Increase of industrial production and creation of new levels of consumption by scientific research made basic structural reform of American society and economy unnecessary. Millikan enjoyed illustrating this theme with a favorite fable. The hilly area of a country was frequently flooded, so the fable goes. Public opinion concerning the best technique of flood prevention was divided into two groups. One faction thought the flooding could be stopped by leveling the hills and filling the valleys. The other faction recommended either lowering the water table or raising the level of all the land. This was a silly fable and certainly did not illustrate what Millikan thought it did. But to Millikan, the economic moral was conclusive: the first group constituted the political and economic reformers; the latter group, the scientists and engineers. Though both groups might be necessary to a dynamic society, Millikan argued, "the second group is less likely to make costly mistakes and more likely to accomplish useful results." Thus the reformers meant well but were dangerously misguided.[37]

The outright redistribution of income would not significantly alter the standard of living of lower classes. The redistribution would have, however, the adverse effect of removing incentive and thereby destroying the "leadership and brains" of America.[38] Millikan did not notice that this pronouncement contradicted the dictum that science promoted altruism.

36. Millikan, "Science and the Standard of Living," *Forum* 99 (March 1938): 175. Millikan, "The New Opportunities in Science," pp. 294–95.

37. Ibid. See also Millikan, "Science and Society," in *Science and Life*, p. 81.

38. Millikan, "New Opportunities in Science," pp. 294–95. He estimated in 1919 that levelling wealth would raise the income of the worker no more than 3 to 10 percent. The basis of this estimate was not explained and the estimate is meaningless. Millikan, "Science and the Standard of Living," p. 175.

Millikan obviously had a poor understanding of economics and his mind was unscientifically closed to reform. But his situation revealed the conservative posture into which science was forced in the 1920s in its attempt to reinforce traditional values and to build a consensus. Millikan's views represented the center of prewar progressive thought. This point can be illustrated most appropriately by brief reference to the prewar economic views of Frederick W. Taylor, father of scientific management.

The underlying assumption of Frederick Taylor's system of scientific management was the identity of the interests of the employer and the employee: "Prosperity for the employer cannot exist through a long term of years unless it is accompanied by prosperity for the employee, and *vice versa.*" From this identity of interests in expanding prosperity, it followed that the most important task of management and the worker was training him to work efficiently to his maximum capabilities. The increase in profits deriving from this efficiency allowed management to increase the incentive pay, thereby bettering the working man's condition.[39]

Taylor explained to the social committee of the House of Representatives in January 1912 that management's and the employee's recognition of their mutual interest required a "mental revolution." Both had to realize that the basic problem in industrial capitalism was not "what may be called the proper division" of profits and wages (the "surplus") or redistribution of wealth, but the enlargement of the surplus itself. "Both sides [must] take their eyes off of the division of the surplus as the all important matter, and together turn their attention toward increasing the size of this surplus until this surplus becomes so large that it is unnecessary to quarrel over how it shall be divided."[40] Taylor thought that the "social problem" of the existence of great poverty beside extravagant wealth would be solved as production

---

39. Frederick W. Taylor, *The Principles of Scientific Management* (New York: Harper & Brothers, 1911), pp. 10, 12. Initiative and incentive were discussed by Taylor on pp. 33–36.

40. Taylor, *Scientific Management* (New York: Harper & Brothers, 1947), pp. 8–12, 27, 28, 29–30.

increased. There was no need for the structural reform of society or the economy. There was likewise no need for labor unions and collective bargaining because these merely accentuated the antagonism between labor and management in the quarrel over the surplus.

In many respects, Taylor's system of scientific management epitomized central progressive attitudes (in contrast to the leftist progressive attitudes of Herbert Croly, for example, which deemphasized individualism and stressed collectivism). The system affirmed the common interest of the worker, management, and middle class in efficient, profitable production, thereby diminishing class strife. It reflected middle-class anxiety over labor unions. It reinforced progressive individualism by its emphasis on the efficient use of individual capabilities. More important, it made industrial capitalism itself progressive. By the expedient of a "mental revolution," a revolution of social and economic structures was avoided.[41]

The popularity of efficiency doctrine in the late progressive period, initiated by the Eastern Rate Case of 1911, and the absorption of efficiency into progressivism provided a traditional value which national scientists in the 1920s could utilize in building a national consensus on science. The war experience of mission research had brought efficiency into the scientific method itself. Thus, efficiency and the increase of wealth by use of scientific method were both values bridging the prewar and postwar periods and scientific and nonscientific sectors of society.

Scientific Education

The reinforcement of individualism and the private creation and free distribution of wealth through the scientific method led directly to the increased significance of scientific education. For it was scientific education which ultimately gave the individual

41. This topic is discussed in detail in Samuel Haber, *Efficiency and Uplift: Scientific Management in the Progressive Era, 1890–1920* (Chicago: University of Chicago Press, 1964) and Milton J. Nadworny, *Scientific Management and the Unions, 1900–1932* (Cambridge, Mass.: Harvard University Press, 1952).

the power to stand against the emery wheel of competition. One of the chief objectives of the popularization of Science Service, for example, was the education of the public. But of course it was education to a point of view rather than to a set of facts. Formal education had to inculcate the scientific method as well as scientific facts.

Before undertaking full-time duties with the National Research Council in 1916 and later the army, Robert Millikan had a long involvement with education. While at the University of Chicago, he taught elementary physics in a secondary school for five years and for fifteen years did assistance teaching of secondary school physics. He taught a course on science education for ten years at the University of Chicago. He wrote textbooks.[42] Because he considered scientific training the best preparation for thinking about social and life questions, Millikan thought the secondary school curriculum should be organized around the natural sciences. Science should be taught so that the student had a basic comprehension and working knowledge of it at graduation. This basic comprehension was required if a large corps of scientists were to be educated on the college level: "The war has demonstrated the value of science, it has waked up some of our leaders to its possibilities, and has emphasized the necessity for thoroughness in scientific training."[43] Scientific education was necessary to give every student, not alone those intending to be scientists, an understanding "of the scientific mode of approach to life's problems, to give . . . some familiarity with at least one field in which the distinction between correct and incorrect or right and wrong is not always blurred and uncertain, to let him see that it is not true that 'one opinion is as good as another.' "[44] No society governed by ballot, rather than by bullet, as Millikan put it, could afford to have an uneducated electorate.[45]

42. Millikan, "The Problem of Science Teaching in the Secondary Schools," *School and Society* 22 (November 21, 1925): 634.

43. Millikan, "The Present Needs of Science Instruction in Secondary Schools," *School Science and Mathematics* 20 (February 1920): 104.

44. Millikan, *Science and the New Civilization*, pp. 46–47.

45. Millikan, "Education and Unemployment," p. 804.

John Merriam subscribed to Millikan's desire for reorganization of secondary education around scientific studies with a strengthening of the scientific studies themselves. Merriam believed it was the educator's responsibility "to bring about a better understanding of the relation between the two great ideals of *construction* and of *service* which are fundamental to right living." For Merriam, the ideal of constructive living was fulfilled by adopting the attitude of scientific research toward life's problems. He echoed the common plea of the 1920s for continuing education (after formal schooling) to keep the democracy's citizenry adequately informed. Democracy demanded that citizens be educated to address "great questions." For this reason, Merriam objected to the increasing professionalization of education. Education based on scientific or research studies should be liberal, that is, be devoted to teaching the student how to ask meaningful questions and how to seek answers. Education should not be vocational training. Education should develop the power of judgment, not concentrate on the transmission of factual information.[46]

John Merriam did not offer specific proposals for the reform of education. Robert Millikan did. Millikan was distressed that in the competition for study time in the high school curriculum, one less course in the natural sciences was advised to provide time in the student's schedule for, say, a required foreign language. At graduation, most students had a less than cursory education in science. Since the intense study of the sciences came late in the school program, the student would be past the age of natural curiosity about the natural world. The remedy for these shortcomings came to Millikan apparently when he had to review the educational record of a Danish student seeking admission to the University of Chicago. The Danish student had begun his intense scientific education a full six years before his graduation. Physics, chemistry, and botany were taught several hours a week each for the six years, with zoology being added the fifth year before

---

46. Merriam, "The Research Spirit," p. 2380. Italics in text. Merriam, "Making a Living—or Living?" p. 2036. See also Merriam, "Spiritual Values and the Constructive Life," p. 2055, and "Reality in Adult Education" (April 1933), *PPA* 4: 2444.

graduation. Science instruction began in the student's early ado-
lescence when the natural curiosity about the physical world was
greatest. This natural interest was nourished by continual contact
with the science until graduation. Because each science was
taught only a few hours each week, the subject material had an
opportunity to be absorbed. Moreover, because it was taught for
six years, it was not quickly forgotten. Another advantage of this
program of studies was that physical science was taught before
mathematics, thus making concrete examples from the former
available to the latter. Elementary physics could be taught with a
minimum of mathematics. [47]

Millikan did not think the content of secondary school courses
had to be changed or particularly that methods of instruction had
to be renovated. To the contrary, he considered organizational
and administrational changes sufficient. [48] The only other criticism which Millikan seriously voiced concerning science instruction was the lack of training in science itself which many teachers
exhibited. Teachers could not be expected to instill an understanding of the scientific method in their students when they
themselves had not had firsthand experience with it. The teachers
should be trained in scientific research, not in science education. [49]

Millikan's views on education went through only one significant change in the interwar period. In response to the depression,
Millikan became convinced that it was impossible to put scientific
education on a mass level. He recommended that many students
should enter vocational and trade schools. The middle class, he
thought, was becoming too large. The exact origin of this concern
is not clear. Apparently, it reflected a decreased faith in democracy which Millikan began to feel in the 1930s. Not everyone, not
the average man, was capable of using the scientific method and
managing American society. [50]

47. Millikan, "The Elimination of Waste in the Teaching of High School Science,"
School and Science 3 (January 29, 1916): 165–68.
48. Millikan, "Science in the Secondary Schools," School Science and Mathematics 17
(May 1917): 381, 386–87; Millikan, "Present Needs of Science Instruction," p. 103.
49. Millikan, "The Problem of Science Teaching," pp. 635, 637.
50. Millikan, "Education and Unemployment," pp. 809–10.

The educational views of Robert Millikan and John Merriam in the 1920s, like their social and economic views, derived from the center of prewar progressive thought. These views were not original contributions to American educational theory, but they did reveal how the ideology of national science involved one of the main concerns of progressivism. As Lawrence A. Cremin has broadly stated, this progressive concern was the reform of American education with the objective of making education "an adjunct to politics in realizing the promise of American life."[51] Millikan's proposed reform of secondary school science instruction was entirely within the boundaries of this objective. The promise of American life was implicit in the title of Millikan's 1919 lecture, "The New Opportunities in Science." And the progressive objective for education certainly encompassed Merriam's desire that education be made an effective preparation for the modern world by centering its curriculum around the scientific or research method.

The individualism of Millikan's and Merriam's social philosophy, moreover, represented the dominant line of postwar progressive pedagogical theory. Child-centered pedagogy emphasized the notion "that each individual has uniquely creative potentialities and that a school in which children are encouraged freely to develop these potentialities is the best guarantee of a larger society truly devoted to human worth and excellence."[52] The national scientists would not have accepted the extreme manifestation of this pedagogical philosophy, the play school, but they would have agreed that the development of the individual's capabilities, through scientific studies, was the central aim of the school.

While one concern of the national scientists was in the center of progressive educational theory, another of their concerns—reinforcing traditional values through scientific education—partook of a conservative response to progressivism. One historian has argued that in the 1920s the conservatives viewed education as the process of training an elite to manage the

51. Lawrence A. Cremin, *The Transformation of the School: Progressivism in American Education, 1876–1957* (New York: Alfred A. Knopf, 1961), p. 88.
52. Ibid., pp. 201–02.

nation.[53] While the national scientists in the twenties did not overtly accept the concept of an elite since this would have conflicted, at least rhetorically, with their endorsement of democracy, they did accept the view that economic and political positions of authority should be occupied by persons qualified by intellect and experience. Slosson and Millikan thought the citizenry ought to be well enough educated to acknowledge and to elect persons qualified to lead the nation. Rush Welter is perhaps inaccurate in thinking that the concept of the business-managerial elite was primarily a conservative response to progressivism, since the concept of the administrative expert who was scientifically trained was central to progressive democracy. The national scientists were able to appeal to the latter concept as much as to the conservative concept.

The Scientific Approach to International Relations

A final, though not central, aspect of the ideology of the national scientists concerned international relations. Their views on internationalism originated in a number of ways. First, at least for Millikan, the solution proposed for domestic problems—the scientific method—was proposed also for the solution of international issues. Second, the war experience was not only one of national science but also one of international science. Finally, the close connection between their views on national and international problems was itself typical of progressivism.

At the core of the scientists' views on nationalism and internationalism was a dilemma that they first encountered in the war. On the one hand, out of the experience of international cooperation in scientific research during the war and from the historical belief that science knew no national boundaries, the national scientists believed that science could promote peacetime international cooperation and the arbitration of international disputes by a rational method of scientific research and judgment. On the

53. Rush Welter, *Popular Education and Democratic Thought in America* (New York: Columbia University Press, 1962), pp. 289–93.

other hand, in their attempt to build a public consensus in favor
of national science, one of the scientists' strongest arguments was
that scientific research would promote America's industrial,
commercial, and military supremacy in international competition.
Without doubt, this latter consideration was derived from the
knowledge that Germany's prewar supremacy in industrial sciences, like organic chemistry, was a factor in that nation's military power. This dilemma was intensified by several factors. Millikan, like many advocates of the League of Nations, thought that
nationalism in international affairs was an irrational and disrup-
tive force. Nevertheless, the national scientists were allied to
precisely those national leaders who represented industrial and
commercial nationalism—Herbert Hoover, Elihu Root, Charles
Evans Hughes, and the industrialists who were interested in
economic penetration of foreign markets. This alliance was forged
in the scientists' appeal for money. Finally, the scientists themselves, although scientific internationalists, were also scientific
nationalists. They were committed to raising American science
from the provincial level to the national level. Indeed, they were
intent on making American science as good as and better in some
fields than European science. The intention was undoubtedly
strengthened by the awareness that a scientific revolution was
underway, the beginning of which American science had missed.
One of Robert Millikan's objectives in taking the headship of the
California Institute of Technology was to raise the level of American pure science. All would agree that at least this of Millikan's
projects was successful.

"The lesson which physics has to teach the world," Robert
Millikan asserted, while the First World War intently battled on,
"is that war can and must be abolished." War could be abolished
if all nations adopted the scientist's method—"the method of
calm, judicial, reasoned action." Millikan's faith in the scientific
method was so pervasive that he did not need to undertake any
extensive reasoning with regard to the causes of war, settlements,
or the conciliation of international disputes. Superficial considera-
tions of complex matters must end in contradictions. Thus Millikan could say, also during the war, that the mastery of nature by

science and the application of scientific method to commerce and industry was the only way to achieve national greatness and "keep in the forefront of progress." It did not apparently occur to Millikan then that the pursuit of national greatness and the applications of science to industry had been contributing causes of the Great War.[54]

Millikan concluded that science had made war too dangerous for war to be a useful instrument of nationalism. The survival value of war had been diminished: "Because of the growth of modern science it is no longer profitable for a country to engage in aggressive war. The risks to its own life are too great, as the last war showed." Millikan did not think these pious phrases were as hollow as they must appear today, of course. Accepting a vague metaphor of natural selection, he thought that science had given peaceful activities an advantage over militarism in the competitive struggle for survival. Unconsciously reiterating Victorian myths, he believed that the politically neutral character of science, the intercommunication and trade between nations, and "the necessity for international understanding" had made war detrimental to national interest.[55]

It is not out of place here to discover what impact the development and use of the atomic bombs in World War II made on Millikan's views. In the late 1920s, Millikan assured the public that atomic energy could not possibly have harmful effects because nature was so constructed to prevent them. And science itself made another war very unlikely after World War I. He had the opportunity to reconsider these matters in the second and revised edition (1947) of his work, *Electrons (+ and −), Protons, Neutrons, Mesotrons, and Cosmic Rays*. Millikan's assumption that atomic energy could not be dangerous had been dogmatic. He had precluded that any discovery should modify the knowledge he held about atoms. The discovery of the neutron and later (1938–39) the discovery of nuclear fission, of course, made his

54. Millikan, "The New Physics," p. 620. Millikan, "Science in the Secondary Schools," p. 380.

55. Millikan, "Science and Modern Life," in *The Creative Intelligence*, p. 157; see also *Science and the New Civilization*, pp. 62–63, 184–85.

knowledge inadequate and exposed his arrogance. He tried to excuse himself by saying that these discoveries could scarcely have been foreseen. But Millikan now held with elaborate scientific justification that controlled fission would never be a source of energy for peaceful purposes (such as the production of electric power). This assertion was made in the belief that no adequate resources of uranium would be discovered and that no inexpensive technique for exploiting uranium would be developed. The atomic bomb itself only reinforced his prewar opinion that science had made war too dangerous: "Time was when war may have served a useful purpose in the development of the race, but with the advent of modern total and global war no thinking man can question that that time is past."[56] It did not occur to him to ask how this lesson could have been so clear after the First World War, if the Second World War had just occurred.

Since war was not a rational decision, Millikan thought, peacekeeping was essentially a task of controlling and dissipating irrationalism. The institutionalization of the scientific method in an international congress, such as the League or the United Nations, should provide an effective means of control. The League embodied "*the objective mode of approach* to international difficulties."[57] This endorsement was expanded after the Second World War to include world government which he thought could inspect and control a ban on atomic bomb manufacture. His vision of the role of science in international affairs was a direct extension of the progressive's vision of the role of science in domestic affairs. The progressives before the war valued the scientific method and the scientist for neutral arbitration between the competing interests of the corporations, the middle class (the "people"), and unionized labor. On the international level, the scientific method became the neutral, or objective, arbitrator between conflicting national claims. The League of Nations was an international "commission," the commission having been the

56. Millikan, *Electrons (+ and −), Protons, Photons, Neutrons, Mesotrons, and Cosmic Rays* (1935; reprint ed., Chicago: University of Chicago Press, 1947), ch. 11.

57. Millikan, "Science and Modern Life," in *The Creative Intelligence*, p. 155. Italics in text.

progressive embodiment of the scientific method. Indeed, Millikan later called world government a "commission."[58]

Toward Scientific Oligarchy

Two central themes ran through the scientific and social values of the ideology of national science. In their endeavor to establish a public consensus for national science, with the objective of perpetuating the cultural unity and the new scientific institutions created in the First World War, the scientists related the values of science derived from the scientific method to the central values of prewar progressivism. And as the coherence of prewar progressive values dissolved in the 1920s, the scientists proposed that science guaranteed the perpetuation of these values. This conservative posture into which the scientists had cast science was parallel to the conservative appearance progressivism itself gained as the new urban liberalism arose.

This conservatism was illustrated sharply by Robert Millikan's response to the depression and Roosevelt's New Deal. In particular, Millikan's belief in the possibility of an educated mass democracy diminished. He proposed, in opposition to the politically democratic formulation of policies to solve the problems of the depression, what he called the method of the scientific jury.

In an article submitted to *Harper's Magazine* but rejected in September 1934, Robert Millikan described the great depression—which had just put Americans through the worst year of economic suffering in their history—as "a [temporary] jam in the social machinery which makes it impossible at the moment to reap the benefit of the success of the scientist and the engineer in creating more wealth, and in thus raising the general level of human well-being."[59] It deeply disturbed him that the Roosevelt administration chose to alleviate this suffering and to repair the social machinery by make-work projects and economic

---

58. Millikan, *Electrons (+ and −)*, pp. 440–41.
59. Millikan, "In the Coming Century" (unpublished ms.); L. F. Hartman, ed., *Harper's Magazine*, to Millikan, September 6, 1934, box 48, Millikan Papers.

and fiscal legislation regulating commercial practices. This was the "patronage system" which destroyed individualism, the private creation and free distribution of wealth.[60]

Popular support for Franklin Roosevelt led Millikan to doubt the efficacy of political democracy. The people could not be well enough educated, even in scientific reasoning, to vote correctly on policies. He made a campaign speech for Hoover in 1932 in which he said that a public educated to decide wisely on policy was "perhaps forever unattainable."[61] The 1932 presidential election was to test whether the people could so vote. Hoover lost the election, of course, and the people resoundingly failed the test. But Millikan's doubt of the people's capacity to make policy decisions, or even to vote for the right leaders, did not originate with Hoover's defeat. As early as 1930, Millikan thought he detected declining public idealism and growing lawlessness sufficient to make the public's responsibility suspect. That same year, John Merriam indicated similar discomfort with political democracy by saying, in a tone of diminishing assurance, that he was "still of the view that democracy is a useful experiment."[62]

Convinced that the scientific method sanctioned only an individualistic society with free economic initiative, Millikan was forced to believe New Deal policies were created either because their creators were scientifically ignorant or because they were attempting to buy votes. He considered it a mistake that policies were based on the social sciences, rather than on the scientific method of the natural sciences. This was a mistake because the social sciences had an almost negligible body of agreed-on facts and many contradictory theories. The situation in the natural sciences was the reverse of this.[63]

Reasserting the primacy of the scientific method for the new economic problems, Millikan described a technique for the dis-

---

60. Millikan, "Science and the Standard of Living," p. 176.
61. Millikan, "Why Vote for Hoover" (unpublished ms.), box 48, Millikan Papers.
62. Millikan, *Science and the New Civilization*, pp. 84–85. Merriam, "The Opportunities of the Federal Government in Research," *PPA* 4: 2471.
63. Millikan, "The Diffusion of Science: The Natural Sciences," *Scientific Monthly* 35 (September 1932): 203–08.

covery of agreed-on social facts on which policy could be based—the "jury method." It was elaborated in two articles, "Science and the Standard of Living" (March 1938) and "New Frontiers in Economic Progress" (June 1938). According to Charles S. Peirce, one of the functions of the experiment was to provide an objective determination of facts ("conceivable effects") on which investigators could agree. Lacking possibility for experimentation in the adaptation of the scientific method of the natural sciences to social problems was compensated for, in progressivism, by the democratic faith in the judgment of the scientifically trained person. If faith in democracy were to diminish, inability to experiment would become a problem. Democratic balloting, each citizen guided by the scientific method which removed the influence of interests, was supposed to provide scientific agreement on the correct policies. Without faith in the scientific educability of the public, a substitute form of experiment had to be devised to reach scientific agreement.

How would a nonscientist, Millikan asked, discover what was fundamental knowledge of a natural science? He would simply ask a scientist the question, " 'Will nine out of ten of the most competent, experienced, and dependable physicists [for example] agree on the answer to the physical question about which I desire knowledge?' " If the scientists agreed, an established fact was known. Millikan thought this method was required in the social sciences to discover fundamental knowledge. A jury of authoritative economists could establish what would and what could not bring national prosperity. Millikan himself had used the jury method—the jury not being named—to discover certain economic facts. For example, he learned that the standard of living could be raised only by increasing industrial production through the application of science.[64]

Millikan's method of the scientific jury was not significant for the specious fundamental knowledge it yielded. Rather, the jury

64. Millikan, "Science and the Standard of Living," pp. 172–73; Millikan, "New Frontiers in Economic Progress: The Fallacies of Marx" (Address before the 26th annual meeting of the U.S. Chamber of Commerce, Washington, D.C., May 2–5, 1938), *Vital Speeches of the Day* 4 (June 15, 1938): 538.

method signified a shift away from the progressive faith that an educated electorate could decide national policies toward the conservative conviction that these policies had to be established and administered by a scientifically trained jury, who were themselves not popularly elected but chosen on the basis of their authority by other scientists. This shift to scientific oligarchy was strikingly revealed in Millikan's speech "Science and Social Justice" (October 1938). All hope for human betterment depended on science. But what, he asked, was social justice? What distinguished the just act from the unjust act? Here Millikan had to move away from the scientific democracy. Not every person was competent to determine what was the just act. Justice was "human betterment as a whole."[65] Definitions of justice derived from humanitarian sentiment or class conflict were equally wrong. Only the scientist was qualified to determine what would be an improvement of man's condition as a whole. Only he would know what action could produce this improvement.

In America, social justice could be achieved only when the scientific jury was instituted in government. Millikan elaborated his plan in a statement that is worth quoting:

> Those in control [of the government] must either themselves be thoroughly trained in the method of the modern, correct attack upon the problems of economics, finance and government, or must at least be willing to choose as their advisers only those who are recognized by their peers in their own fields as the ablest, most high-minded, most competent men in those fields, irrespective of whether they speak the shibboleths of the men in power or not. That alone constitutes the scientific approach to the problem of government.[66]

Millikan was thus finally brought to the conservative position to which many of the prewar scientific progressives, like Walter

65. Millikan, "Science and Social Justice: 'A Stupendous Amount of Woefully Crooked Thinking' " (Speech before the N.Y. *Herald Tribune* Forum, October 25–28, 1938), *Vital Speeches of the Day* 5 (December 1, 1938): 99.
66. Ibid., p. 100.

Lippmann, had come in the 1920s. The faith in democracy and reform had diminished, but not the faith in science.[67]

The intellectual shift in Millikan completed a cycle which led to an ultimate contradiction of the ideology of national science formulated in the 1920s. The cycle began with the pre–World War I separation of the scientific and nonscientific cultures. The cycle went through the war, when the scientists came to recognize the benefits of a united culture, into the 1920s, when they attempted to formulate an ideology unifying scientific and social values and to establish a public consensus for national science. The cycle ended in the 1930s with Millikan's position that the democratic masses were not scientifically educable, could not vote scientifically for national policies, and that, therefore, only the scientists were competent to use the scientific method. The scientists were isolated once again.

67. This transformation is discussed in Kaplan, "Social Engineers as Saviors."

(7) THE FAILURE OF THE IDEOLOGY, 1926-1930

The endeavor of the scientists to build a public consensus on the importance of national science culminated in the campaign for the National Research Endowment, "A National Fund for the Support of Research in Pure Science," from 1926 through 1930. Eminent politicians, businessmen, and scientists appealed to the public and to corporations for twenty million dollars which was to be dispersed to scientists engaged in research having no necessary commercial value. Trustees of the National Research Endowment included Herbert Hoover, Elihu Root, Charles Evans Hughes, Andrew W. Mellon, and Edward M. House among political and business leaders, and George E. Hale, Robert A. Millikan, and John C. Merriam among scientists. The endowment was to be held by the National Academy of Sciences and administered by the trustees, chosen by the National Academy. Campaign publicity was intense and continuing. Industrial corporations were solicited for contributions. The appeal for funds was based on the ideology of science.

Despite the prestigious support for the endowment, by the fall of 1930 the campaign had unmistakably failed in its objectives.[1] This failure was not simply the inability to secure pledges of twenty million dollars. The scientists had been unable, more

---

1. Surveys of the history of the National Research Endowment are available in A. Hunter Dupree, *Science in the Federal Government: A History of Policies and Activities to 1940* (New York: Harper & Row, Torchbooks, 1964), pp. 340–43; and Helen Wright, *Explorer of the Universe: A Biography of George Ellery Hale* (New York: E. P. Dutton, 1966), pp. 365–70. Dupree's account contains several historical errors which have been corrected in Wright. Neither has considered the topic in the broader and ideological framework that defines my approach.

importantly, to maintain the independence of pure science from industrial capitalism and to secure a primary relationship with the general public like that of the expert in the progressive period who staffed the commissions and protected the middle class interest. Freedom from industrial economics and unity with the lay public had been central values in the scientists' ideology. The subversion of these values during the campaign indicated the scientists' failure to link pure science with the progressive faith in the people and the conservative shift in their values.

The National Research Endowment

The public announcement of the campaign to raise a research endowment was made in February 1926. But the endowment plan itself was the solution for problems that originated at the end of the world war. One problem was organizing and financing the postwar National Research Council. Another was the isolation of pure science in a peacetime culture, which contrasted with the public acceptance of applied science and engineering. The precise character of the solution to these problems—a public endowment—resulted from placing the support of pure science within the framework of the ideology of national science.

The National Research Council emerged from the First World War under the leadership of a coherent group of scientists who wanted the council's national unification of the sciences to continue in the peace. The first obstacle to this goal was financing because the money from the president's special fund which had supported the council's work would no longer be available. This problem was solved by private philanthropies (the Carnegie Corporation, the Rockefeller Foundation, and the General Education Board). Another obstacle was finding adequate support for new programs such as the National Research fellowships and the national laboratories for physical and chemical research. The fellowships were eventually financed by the Rockefeller Foundation and the General Education Board (which provided for medical fellowships). The research laboratories were never supported, in part because the scientists themselves could not decide precisely what research institutions they wanted.

Numerous other research proposals and financial schemes passed among the national scientists in these immediate postwar years, stimulating consideration of the place of science in America, the relationship between federal, state, and private science, and the relationship between the pure and applied sciences. It was in this milieu that George E. Hale, for instance, wrote to John Merriam in 1920 concerning a proposal, originating in a council committee, for governmental aid to private research. The proposal raised the question of whether the National Research Council would come under governmental supervision, which Hale did not want. But there was a further problem with securing governmental aid. The national scientists were encountering difficulty in gaining the government's acknowledgment of their presence and purpose. The engineers, Hale thought, had ingratiated themselves with Congress and the executive administration to the exclusion of the basic scientists. They had, moreover, "been very clever in identifying [Herbert] Hoover with their organization, and by this and other means they seem to be establishing themselves very firmly in Washington." Hale hoped that Hoover could be persuaded to help the scientists make the Research Council the authoritative national voice on matters of research.[2]

Hale's letter to Merriam synthesized the difficulties facing the pure scientists into one problem—obtaining public recognition of the distinctiveness and importance of pure science. Hale perceived that some of the public awareness and support of organized engineers derived from Herbert Hoover's reputation as an engineer and his representation of engineering in governmental service. Hale was to conceive of a national endowment not only

2. George Ellery Hale to John Campbell Merriam, November 27, 1920, box 87, John Campbell Merriam Papers, Manuscript Division, Library of Congress (cited hereafter as Merriam Papers). With regard to the efforts to win Hoover, see also Robert M. Yerkes (chairman, Research Information Service) to John Merriam, February 17, 1921, box 225, Merriam Papers.

as a means of obtaining research money, but as a means of gaining public recognition of pure science. The image and influence of Hoover and other nationally famous leaders were to be identified with pure science by the endowment campaign.

In the spring of 1922 Hale proposed to Millikan, Merriam, and Vernon Kellogg a scheme for a public subscription, national research fund to finance scientific research. This suggestion could well have been made at the annual meeting of the National Academy of Sciences, April 24–26, when all were together, except Kellogg who was not a member of the academy, but who was in Washington for his duties as secretary of the Research Council. The proposal was subsequently discussed between Hale and Millikan while both men were in Oxford, England, in the summer, and then between Millikan, John J. Carty (National Academy member and vice-president of American Telephone and Telegraph Co.), Gano Dunn (president of J. G. White Engineering Co.), Vernon Kellogg, and Frank B. Jewett (National Academy member and president of the Bell Telephone Laboratories in New York, a subsidiary of A.T.&T., of which he was also a vice-president) in September 1922.[3]

Hale may have made the formal proposal to his friends for the endowment fund in the spring of 1922, but certainly he had given consideration to the idea before then. Several policy struggles within the National Research Council after the war could have brought the idea for the fund to Hale's mind. It is important to understand that the concept of the fund came out of the same postwar ambience that precipitated the effort of the national scientists to formulate and to proselytize an ideology of national science. One such policy struggle concerned the proposal for a national research laboratory and the research fellowships. The dispute over the laboratory was still smoldering in 1923 and

3. Helen Wright begins her discussion of the endowment at this date also; *Explorer of the Universe*, p. 365. The date of spring 1922 is established in a letter from Millikan to Hale, September 29, 1922, box 29, George Ellery Hale Papers, The Carnegie Institution of Washington and the California Institute of Technology, Pasadena, California (cited hereafter as Hale Papers). My reservations about the date are in the text. The roll at the academy meeting is in the *Report of the National Academy of Sciences for the Year 1922* (Washington, D.C.: Government Printing Office, 1923), p. 8.

kept in Hale's mind the problem of research support.

The other policy struggle concerning the National Research Council scientists in these years was over the relationship of the council's Division of Engineering to the United Engineering Societies and the Engineering Foundation. The connection between the council and the foundation had been close since 1916 when the council was established. The Engineering Foundation had financed the first year's operations of the Research Council. Organized in 1919, the engineering division maintained close relations with the foundation, receiving grants from it for research and maintaining offices in the United Engineering Societies building in New York City. The functions of the two engineering institutions overlapped because the foundation had been set up by Ambrose Swasey to finance engineering research.[4] In 1922 the foundation made personnel changes in an effort to move more vigorously toward the goal for which it was operated. There was, consequently, some sentiment within the Research Council to abolish the engineering division, relinquishing engineering research entirely to the foundation.[5] At the same time, other opinion in the council wanted to assume outright control over the foundation. Of course Swasey and the foundation board were not agreeable to this.[6] By early 1923, this policy struggle had been worked through, with the relationship between the Division of Engineering and the Engineering Foundation remaining in effect much what it had been.[7]

4. On the first formal relationship between the Division of Engineering and the Engineering Foundation, see *Report of the National Academy of Sciences for the Year 1919* (Washington, D.C.: Government Printing Office, 1920), pp. 72, 86–87. For the division's work, see the succeeding report of the National Research Council in the academy's report, 1922, pp. 24–25, 50–55.

5. This situation was discussed in Millikan to Hale, September 29, 1922, box 29, Hale Papers. Vernon Kellogg, for one, wanted to abolish the division; ibid.

6. Hale to Millikan, November 5, 1922, box 24, Robert A. Millikan Papers, California Institute of Technology Archives (cited hereafter as Millikan Papers); Millikan to Merriam, November 24, 1922, box 125, Merriam Papers; Millikan to Kellogg, November 27, 1922, ibid.; Merriam to Millikan, December 2, 1922, ibid.; Hale to Gano Dunn (second vice-president of the executive board, National Academy), December 3, 1922, box 87, Merriam Papers; Merriam to Hale, January 29, 1923, ibid.

7. The reorganization is mentioned in the "Report of the National Research Council for the Year July 1, 1922, to June 30, 1923," in *Report of the National Academy of Sciences for the Year 1923* (Washington, D.C.: Government Printing Office, 1924), p. 40; "Report

These problems with the Engineering Foundation probably brought to Hale's mind the scheme for an endowed, scientific research foundation. Such a scheme would have been in line with his unsuccessful prewar reform of the National Academy and with the abortive postwar plan for a national research laboratory. The idea of an endowed foundation was before him in the policy struggle and certainly was not incompatible with his long experience with the Carnegie Institution.

Hale thought the plan for a research endowment, as well as the solution of the council's problems, required a radical change in the stance of the National Academy toward the public. The academy home building, to be completed in 1924, would allow the academy to take a new public role as Hale had wanted in 1913. The building would be a "Temple of Science," as Gano Dunn was to say at its dedication in April 1924, not only to the scientists but to all Americans.[8] Weekly meetings at the building could draw the public. Science Service could be utilized to bring the meetings and publications of the academy to many more people.[9] The building was to be more than just an administrative headquarters.

In Hale's plan, the National Academy would dominate any national research endowment to be administered jointly by the academy and the Research Council.[10] If the research fund were only to disburse grants to qualified scientists, then the Research Council would have been the obvious administrative agency. If, however, the purpose of the fund was also to gain public recognition of pure science, then the research foundation would have to be identified with the authoritative voice of national science, the National Academy of Sciences. The foundation would be, at the same time, a phase of the academy's new stance toward the public, signified by the building.

---

for the Year Ended February 9, 1922," *The Engineering Foundation* (New York: United Engineering Society, 1922), p. 8; and in "Report for the Year Ended February 8, 1923," ibid., p. 24. See also Millikan to Kellogg, November 27, 1922, box 125, Merriam Papers.

8. Wright, *Explorer of the Universe*, p. 316. A typographical error in dating appears in Wright's account; the date "March 29, 1922" (line 21, p. 345) should read "March 29, 1923".

9. Hale to Merriam, November 17, 1922, box 87, Merriam Papers.

10. Hale to Millikan, November 5, 1922, box 24, Millikan Papers.

Hale's reasoning appeared in his mediation of the differing opinions over the proposed research fund within the "inner group" of national scientists—those who dominated the National Academy and Research Council, the Engineering Foundation, the California Institute of Technology, the Carnegie Institution, whose influence moved into the Carnegie Corporation and the Rockefeller Foundation, and who spoke for science in America. When Millikan approached Kellogg, Carty, and Dunn in September 1922 about the foundation scheme, Kellogg supposed that the funds could be administered by the National Research Council, as it was constituted, for research projects the council had already set up. The others thought that a new committee would have to be established for administration and new research projects developed. All except Kellogg wanted the National Academy to dominate the administering committee. (The reader will recall that Kellogg was not an academy member and was secretary of the Research Council.) Merriam wanted the Research Council to have nothing to do with the foundation. Despite the weight of opinion against Kellogg's views, Kellogg could not be ignored because of his close connections with the Rockefeller Foundation. He had already aroused the interest of George Vincent and Raymond Fosdick, trustees of the foundation, in giving large sums for scientific research. If these funds were to be secured, Kellogg's good graces had to be cultivated. Hale was not opposed, of course, to the Research Council obtaining Rockefeller money for its current research. But he wanted a separate and "larger" research foundation in the name of the academy. This larger fund should have publicly known personalities like Hoover, Root, and Mellon as trustees to identify it.[11] This last consideration of Hale's merged the function of the proposed research foundation with that of a radicalized National Academy—the

11. Hale called it the "inner group of half-a-dozen men" in his letter, Hale to Merriam, November 17, 1922, box 87, Merriam Papers. The group included Merriam, Millikan, Kellogg, Carty, Dunn, and Hale. If Hale did not number himself among the half dozen, probably he would have included his close friend at the institute, the chemist Arthur A. Noyes. Millikan to Hale, September 29, 1922, box 29, Hale Papers; Hale to Millikan, November 5, 1922, box 24, Millikan Papers. Hale remembered that he had talked to Root in 1922 about the research fund, in Hale to Root, October 15, 1925, box 25, Hale Papers.

inculcation in the public mind of the values, method, and national significance of research in pure science.

Despite the prolonged initial conferences, the scheme for the research endowment was only to exist in discussion in 1923 and 1924. In Pasadena, California, in mid–1923, Hale had frequent conversations with Arthur Noyes, a chemist and personal friend, and Robert Millikan over the proposal. The triumvirate of the California Institute thought the foundation should be announced at the upcoming dedication of the National Academy building. Hale consulted Hoover and Root who urged him to initiate the foundation as soon as possible. Hale hoped to excite public interest in research in the same way as Hoover's work for European relief after the war had its interest in humanitarianism. He was sure "from talks with [Hoover] that he will know how to accomplish this."[12] The magic public reputation of Hoover was to be transferred from engineering to basic science. Merriam had occasional second thoughts about the proposal. He feared that the research fund would serve merely to divert the flow of money from the universities to itself. Millikan and Noyes argued with him that the research endowment would draw funds otherwise unavailable for research.[13]

Finally, at the annual meeting of the Executive Council of the National Academy on April 26, 1925, the Committee on Additional Funds for Research was chosen to prepare a report and recommendations for a national research fund. Hale was chairman and Hoover, Millikan, Andrew Mellon, Gano Dunn, J. J. Carty, Thomas Hunt Morgan (the geneticist), Kellogg, and William H. Welch were members.[14] In May, Hale held a conference in New York on the proposed endowment.[15] The committee presented their report to the autumn meeting of the council of the National Academy on November 8, 1925. The report, written

12. Hale to Merriam, January 12, 1924, box 87, Merriam Papers.
13. Millikan to Hale, March 24, 1924, box 29, Hale Papers.
14. Hale et al., "Report of the Committee on Additional Funds for Research," November 8, 1925, box 63, Hale Papers; David White, home secretary of the National Academy of Sciences, "Minutes—Council Meeting, November 8, 1925," box 53, Hale Papers. These documents contain information about the previous annual meeting of the council.
15. Wright, *Explorer of the Universe*, p. 365; see also citations in note 16.

primarily by Hale, with the aid of Dunn, Carty, Frank B. Jewett, and of course Millikan and Merriam, was accepted.[16] The following day the National Academy formed a special board of trustees to collect and to administer the National Research Endowment, as it was designated for the next two years. The National Research Council was not officially a part of the scheme; the academy alone controlled the endowment.[17]

Shortly thereafter at a meeting of the trustees, the decision was made to approach the corporations for large initial pledges toward the proposed twenty-million-dollar endowment. Hale was to approach Julius Rosenwald of Sears, Roebuck, just after Christmas (of 1925). John J. Carty was to lead American Telephone and Telegraph into the fund, then to solicit General Electric, in the first months of 1926. Herbert Hoover was to make powerful speeches in late 1925, asserting that industrial and economic progress depended on research in pure science. Early in the next year, Hoover was to ask the electrical manufacturing companies for pledges.[18] Plans were laid for Elihu Root to approach United States Steel, Hoover to take on the railroads, and Charles Evans Hughes to be consulted about possible contributions from the insurance companies.[19] Very few pledges were made; those that

16. My conclusion regarding authorship and consultation is inferred from John J. Carty to Hale, May 6, 1925, box 10, Hale Papers; Gano Dunn to Hale, May 6, 1925, box 14, Hale Papers; Hale to Merriam, October 13, 1925, box 87, Merriam Papers.

17. See the citations in note 14 above, and *Report of the National Academy of Sciences: Fiscal Year 1925–1926* (Washington, D.C.: Government Printing Office, 1927), p. 116. The original trustees were Herbert Hoover, chairman, Andrew W. Mellon, Charles Evans Hughes, Edward M. House, John W. Davis, Julius Rosenwald, Owen D. Young, Henry M. Robinson, Felix Warburg, Henry S. Pritchett, Cameron Forbes, Albert A. Michelson, John C. Merriam, Robert A. Millikan, Gano Dunn, Vernon Kellogg, William H. Welch, Thomas H. Morgan, John J. Carty, Simon Flexner, Oswald Veblen, James H. Breasted, Lewis R. Jones, Arthur B. Lamb, and George E. Hale; ibid.

18. Wright, *Explorer of the Universe*, has a fine account of Hale's meeting with Rosenwald, pp. 366–68. Hale to Carty, February 4, 1926, box 10, Hale Papers. Dupree, *Science in the Federal Government*, p. 341. Hoover to Hale, February 8, 1926, box 22, Hale Papers.

19. Millikan to Root, April 9, 1926, box 141, Elihu Root Papers, Manuscript Division, Library of Congress (cited hereafter as Root Papers); Millikan to Hoover, April 9, 1926, box 8, Millikan Papers; Carty to Hale, September 9, 1926, box 10, Hale Papers. Hoover to Millikan, July 2, 1926, box 8, Millikan Papers. Carty to Hale, September 28, 1926, box 10, Hale Papers. The effort of Hughes on behalf of the National Research Endowment with regard to the insurance companies was unintentionally ironic. Hughes had been elected a

were made were conditional on a subscription to the endowment
by United States Steel or on the full subscription of the endowment. Although U.S. Steel never made the entire contribution for
which it was first asked, in the late spring of 1929, the corporation did make a pledge of one hundred thousand dollars a year
for ten years. The attempts to secure large contributions from
corporations early in the campaign followed Hale's theory of
1924 that the general public might not contribute to the endowment until impressive examples had been set for them and the
success of the project seemed assured.[20] The campaign was unable, however, to obtain committed corporate subscriptions before the public announcement of the endowment.

National publicity of the National Research Endowment campaign was widespread and intense. Beginning with comment
about Secretary of Commerce Hoover's speeches at the end of
the year and with the official announcement of the endowment in
early February 1926,[21] public interest was appreciably strong for
half a year, apparently remaining high until the depression. The
Department of Commerce kept a clipping file which indicated
the scope of attention in the early months of the subscription
campaign. The following figures for clippings of newspaper stories received at the department were estimated to be only ten to
twelve percent of the total number of newspapers carrying the
news. The campaign feature release, "Plan Greater Extension of
Research," of early February 1926, appeared in at least twenty
newspapers, including the Jamestown, New York, *Journal* and the
Marietta, Ohio, *Times*. "Elihu Root Declares Progress Depends,"
of January–February 1926, was carried by at least ten papers,
among which were the Astoria, Oregon, *Budget*, and the Fairfield, Iowa, *Ledger*. "Workers in Pure Science Field Need Support," with the photograph of George E. Hale, also of

progressive governor of New York before the war on the basis of his exposures of dealings
between insurance companies and bankers.

20. Hale to Merriam, January 12, 1924, box 87, Merriam Papers.

21. "Say Pure Science Lags in America," *New York Times*, February 1, 1926, p. 7. My
choice of this date as the opening of the official public campaign is arbitrary but, to my
evidence, as accurate as ascertainable.

January–February 1926, was carried by at least six papers, including the Red Wing, Minnesota, *Republican*. Clippings were received from forty-six papers carrying Hoover's speech asserting America's lag in pure science research. News stories about the endowment campaign were published by at least fifty-one newspapers. More important, editorials concerning the campaign were published in at least 304 papers.[22] Science Service, magazines, and radio added to the public exposure of the fund drive.

Soliciting United States Steel

The negotiations between the national scientists and United States Steel Corporation from the spring 1926 to the spring 1929 illustrate the general approach to the corporations. The reason that U.S. Steel finally gave in to the scientists' request was significant because the contributions of other industries were contingent upon a subscription by the giant corporation. More important, however, the solicitation of U.S. Steel was a key event initiating and revealing a subtle shift of the scientists' values toward conservatism.

Early in the week following Sunday, April 4, 1926, Elihu Root and Robert Millikan had a conference in New York with Elbert Gary, chairman of the board of U.S. Steel. Although, apparently, there is no record of their conversation, it can fairly be assumed that Root explained the endowment plan from the legal and organizational side to Judge Gary and that Millikan explained how indebted the steel industry's progress was to pure science. Gary was not convinced. He worried over the apparent legal obstacles to investing stockholders' money in a nonprofit endowment. And Millikan did not persuade him of "the direct and intimate dependence of the modern steel industry on the results of pure science research."[23]

---

22. R. C. Mayer of the office of the secretary of commerce to Hale, February 29, 1926, box 22, Hale Papers. All the clippings figures are from the report accompanying this letter.

23. Frank B. Jewett to Hoover, April 9, 1926, box 8, Millikan Papers.

To meet Gary's objections, Root obtained alterations in the legal trust by which the National Academy was to hold funds.[24] Millikan armed him with a three-page statement on "The Indebtedness of the Steel Industry to Pure Science," written by John Johnston, professor of physical chemistry at Yale.[25] Root apparently saw Judge Gary the following week but to no avail.[26] In September, John J. Carty suggested that he appear before the executive committee of U.S. Steel to make the scientists' plea once again. He hoped that detailed arguments based on the contribution of pure science research to the metallurgy of iron and steel would be effective.[27] But still no subscription from U.S. Steel was forthcoming.

In April 1927, the continual appeals of the scientists to the steel corporation had partial success. A meeting with the Finance Committee of U.S. Steel was arranged for the fifteenth. But an ambiguous factor had appeared. U.S. Steel wanted to make Robert Millikan a vice-president of the corporation in charge of basic research. At a meeting of steel officials earlier that spring, Gary, J. P. Morgan (the son of the founder of the corporation), and others, when the scientists were not present, it was decided that U.S. Steel "must" have Millikan. Millikan would be given unlimited funds. The corporate leaders argued that this arrangement would "advance science enormously."[28] Before he came to New York for the April 15 meeting, Millikan apparently had no intimation that U.S. Steel was going to try to buy him. Judge Gary described the offer over lunch on April 3. Shortly after,

24. "Suggested Modification of the Declaration of Trust, As Proposed by Carty, Millikan, Dunn, Hale, Noyes and Jewett," in *Documents Submitted in Connection with the Request of the Trustees (of the National Research Fund) for a Contribution of $200,000 per Annum for Ten Years from the United States Steel Corporation* (April 1927), box 193, Root Papers.

25. Millikan to Root, April 9, 1926, box 141, Root Papers. At this time, Millikan was delivering lectures at Cornell University and was thus available for these conferences in New York City.

26. The only information I have is that Root was expected to confer with Gary this second time the next week; Jewett to Hoover, April 9, 1926, box 8, Millikan Papers. Whether he actually did is another question. In any event, U.S. Steel did not contribute money this year, so the meeting was unsuccessful.

27. Carty to Hale, September 29, 1926, box 10, Hale Papers.

28. George Ellery Hale to Evelina Hale, April 4, 1927, box 81, Hale Papers.

Millikan hastily called a meeting of the national scientists to discuss the new situation. They feared this offer might be a premonition of refusal to subscribe to the research fund. The next morning at the office of Frank B. Jewett, Millikan, Dunn, Carty, Jewett, and Hale met with a representative of the steel corporation, who repeated and persuasively argued for the offer to Millikan. But Millikan refused; he would not leave the California Institute of Technology. Instead, he made the counteroffer to U.S. Steel that he would take a nonsalaried, advisory position if the corporation would get another person for research director.[29] Later it was agreed that Millikan, together with Jewett and a representative of U.S. Steel, would recommend a suitable director of research.

The "epoch-making" confrontation between the Finance Committee of U.S. Steel and the national scientists over the National Research Fund (as it had now been renamed) took place on April 15, 1927. U.S. Steel was represented by Gary, Morgan, Governor Nathan L. Miller, general counsel to the corporation, and others; the national scientists were represented by Herbert Hoover, Elihu Root, Charles Evans Hughes, Frank B. Jewett, J. J. Carty, Gano Dunn, Millikan, and Hale. Gary opened the meeting. Root then argued for the fund. He sought to satisfy the corporation officers of the legality of a subscription. Root "wound up by saying that it was not only the legal right but the clear *duty* of the Steel Corporation to contribute to [the] Fund."[30] J.J. Carty, Millikan, then Hale spoke. Finally, Gano Dunn outlined the organization and procedures of the National Research Fund. Dunn made quite explicit the importance of U.S. Steel's solicited contribution. After a year and a half of campaigning, the fund had received only two subscriptions. A.T.&T., pushed by Carty, had made a pledge of two hundred thousand dollars a year for ten years, conditional on the full subscription of two million dol-

29. George Ellery Hale to Evelina Hale, April 6, 1927, ibid.

30. J. J. Carty called the meeting "epoch-making"; George Ellery Hale to Evelina Hale, April 17, 1927, ibid. The name was changed to "fund" from "endowment" before the meeting with U.S. Steel. The change represented an accommodation to Julius Rosenwald and Judge Gary who disapproved of invested endowments. Subscriptions would be made in yearly installments and so disbursed.

lars a year for ten years by other donors as well. Julius Rosenwald, whom Hale had met, pledged one hundred thousand dollars a year for ten years, also similarly conditional. Millikan had
visited George Eastman in Rochester, New York, before the steel
corporation meeting to request a contribution. But Eastman
would make no decision. Other potential contributors were similarly hanging back. A dramatic gift from U.S. Steel could push
the fund to its goal.[31]

Elbert Gary remained unconvinced of the legality of a subscription by the corporation. He asked for the opinion of Charles
E. Hughes. Hughes made an impressive reply. Hale wrote to his
wife, "I have never heard a more masterly statement than
Hughes made, beginning in a way that made us fear he would be
against us, but winding up emphatically in our favor." Gary's
objections were apparently muted. Miller seemed likewise
swayed by Hughes's speech.[32] In view of the legal research done
by Root in April 1926 to establish beyond doubt the legality of
the requested contribution to the fund,[33] however, it can be
doubted that Gary's and Miller's reticence to endorse the fund
arose from legal obstacles. In the meetings of Gary and the
national scientists in 1926, the scientists' own impression then
had been that the demonstration of steel's indebtedness to pure
science had not been rigorous enough. This was why Root asked
Millikan for scientific advice and why, in September, Carty
sought information on metallurgy before another meeting with
U.S. Steel. Nor did the meeting with the Finance Committee on
April 15, 1927, now produce any more rigorous demonstration of
steel's self-interest in pure science. Hale spoke vaguely to the
committee about "practical results from improbable sources."
He mentioned the impact of Einstein's theories on the study of
the constitution of matter. J. J. Carty had told the Finance
Committee that the fund trustees would place no restrictions on

31. Gano Dunn, "Notes on the Organization and Procedure of the National Research
Fund as Presented to the Finance Committee of the U.S. Steel Corporation, April 15,
1927," in *Documents Submitted*. George Ellery Hale to Evelina Hale, April 14, 1927, box
81, Hale Papers.
32. George Ellery Hale to Evelina Hale, April 17, 1927, box 81, Hale Papers.
33. Root to Hoover, April 12, 1926, in *Documents Submitted*.

the fields to be supported.[34] None of these statements would have reassured the steel magnates that they would receive any return, no matter how indirect, even in the long run, from their endowment. The persistent legal objections of Gary were most likely a cover for his disbelief that U.S. Steel would benefit from a subscription and a protection for his inability to argue on scientific grounds with the national scientists. There is no evidence that he ever believed in their ideology.

No decision on the requested contribution was made at the meeting of April 15. The scientists left the meeting with high optimism, but U.S. Steel decided not to subscribe in 1927. U.S. Steel's decision to contribute did not come until April 23, 1929— two years after the epoch-making confrontation and three years after they were first solicited. The pledge was for one hundred thousand dollars a year for ten years, only half the original request.[35] Why did U.S. Steel finally make a pledge? The answer is important in judging the effectiveness of the campaign for the National Research Fund. The steel magnates never accepted the scientists' case. The weighting factors which brought the pledge seem to have been the friendship established in 1927 between the scientists, particularly Millikan, and J. P. Morgan, and Millikan's influence behind a plea in 1929 that the campaign was about to collapse unless U.S. Steel gave.

Morgan apparently did not care about the fund in 1927 but wanted the services and prestige of the Nobel Prize physicist. The action of Millikan in declining a vice-presidency as advisor must have impressed and indebted Morgan. In the days following this action, before the epoch-making confrontation, Millikan saw Morgan at least one more time.[36] What conversations were held

34. Hale, "Notes for Talk to Finance Com. of Steel Corporation, April 15, 1927," box 63, Hale Papers. "Statement for General Carty on the Proposed Methods of Administration and Objectives of the National Research Endowment," April 11, 1927 (read by Carty at the U.S. Steel meeting of April 15, 1927, but apparently written for him by Gano Dunn), box 14, Hale Papers.

35. Hale to Max Mason of the Rockefeller Foundation, June 25, 1929, box 63, Hale Papers.

36. George Ellery Hale to Evelina Hale, April 4, 1927, April 6, 1927, box 81, Hale Papers.

cannot be known, but Millikan's personal charm when soliciting funds was legendary in scientific circles and certainly must have cemented his influence with Morgan.[37] When the April 15 meeting took place, Hale could write later to his wife, "Morgan, we know, is strongly with us."[38] Millikan's influence inside the steel corporation, in general, also grew with his position as nonsalaried advisor. Though the Finance Committee was reluctant to contribute, the scientists continued their assault.[39] Finally on April 23, 1929, Millikan and Carty appeared before U.S. Steel's Finance Committee. Root, who was ill, and Hughes could not attend. Elbert Gary had died in August 1927. Millikan, pulling on all his influence, strongly requested half the amount requested in 1927. The fund was going to collapse, he said, and only U.S. Steel could save it. The Finance Committee made its contingent pledge.[40]

The strength of personal influence, rather than the strength of the scientists' campaign, enlightened self-interest of the corporate leaders, or corporate and public acceptance of the scientists' ideology stood behind most of the subscriptions which the scientists secured. This fact is borne out when the few pledges other than U.S. Steel's are examined. American Telephone and Telegraph, which made the first pledge, was led into the fund by national scientists, J. J. Carty and Frank B. Jewett, who were vice-presidents of the company. Millikan himself had close connections with Bell Laboratories and the parent company. In 1910 and 1911, Millikan sent several of his students to Bell labs upon Jewett's request to help develop the transcontinental telephone

37. The legend began in 1919 when Millikan helped to secure the $185,000 needed to purchase land in Washington on which to build the home of the National Academy of Sciences. See Wright, *Explorer of the Universe*, p. 312, on the land procurement. Millikan's legend grew by his ability to build the endowment of the California Institute of Technology, and in 1929, by his efforts to secure funds for the research fund.

38. George Ellery Hale to Evelina Hale, April 17, 1927, box 81, Hale Papers.

39. Millikan and Hale to Root, October 21, 1927, box 142, Root Papers; Millikan to Root, June 26, 1928, box 39, Millikan Papers.

40. Millikan to Carty, September 21, 1929, box 10, Hale Papers. The decision to scale down the fund from a proposed twenty million dollars over ten years to ten million dollars over ten years was suggested by Hoover in late 1927 or early 1928 and later accepted in a form modified by Gano Dunn, ibid. Also J. J. Carty to E. A. Clarke, secretary of the American Iron and Steel Institute, June 22, 1929, box 10, Hale Papers.

repeater. (Jewett had earned his Ph.D. in physics under Millikan at the University of Chicago.) And from 1919 through 1925, Millikan was engaged in patent litigation for A.T.&T. against General Electric over pure-electron discharge. The suit ended in favor of A.T.&T.[41]

The contribution of U.S. Steel in 1929 prodded a few other reluctant donors. The American Iron and Steel Institute, which was dominated by U.S. Steel, voted (May 24, 1929) to match U.S. Steel's gift, to be paid when that corporation paid. Millikan traveled to Rochester to persuade his old friend George Eastman to contribute one hundred thousand dollars a year for five years. (Eastman had made a conditional pledge earlier but Millikan had to persuade him to renegotiate it if it were still to be valid.) Then Millikan stopped in Chicago where he gained a pledge from Julius Rosenwald, who always had serious reservations about endowments and who was hostile to Hale in 1926.[42] Added to the pledges from the electrical manufacturers and the National Electric Lighting Association which had been initially approached by Hoover, the fund by July 1929 had pledges of nine hundred thousand dollars a year for five years, contingent upon securing total pledges of one million dollars a year. The scientists made an urgent plea to the Rockefeller Foundation which had financed the research fellowships.[43] Its officials from the first had been in favor of giving research funds. And Vernon Kellogg, who had the greatest influence of the national scientists with the foundation, was the executive secretary of the National Research Fund. The foundation answered the plea, of course, even contributing an extra twenty-five thousand dollars a year to compensate for the failure of one of the steel contributors to make its pledges.[44]

41. Robert A. Millikan, *Autobiography* (New York: Prentice-Hall, 1950), pp. 116–23. This is discussed at greater length in Millikan's "Address Delivered over the Radio in Support of the Campaign of the University of Chicago, Dictated March 21, 1925," box 48, Hale Papers.

42. Hale to Max Mason, June 25, 1929, box 63, Hale Papers; Millikan to Carty, September 12, 1929, box 10, Hale Papers; Carty to Gano Dunn, October 9, 1929, box 8, Millikan Papers.

43. Hale to Mason, June 25, 1929, box 63, Hale Papers.

44. In 1929, Raymond Fosdick of the Rockefeller Foundation, as well as Max Mason, carried a high interest in the fund; Millikan to Kellogg, May 6, 1929, box 8, Millikan

Having met their obligations, the national scientists decided to implement the fund as quickly as possible. Scientists across the country were impatiently and threateningly waiting for grants, initial decisions about which had been already made. For example, three-year grants of two thousand dollars a year had been suggested for P. W. Bridgman and Arthur Holly Compton, physicists at Harvard and Princeton, and George D. Birkhoff and Norbert Wiener, mathematicians at Harvard and the Massachusetts Institute of Technology.[45] In August and September 1930, notices were sent to the contributors that the fund would begin operations on October 1, 1930, and for that reason pledge payments should be considered due that date. Only $379,000 of the pledged $1 million was received. The National Electric Lighting Association and Julius Rosenwald did not attempt to meet their pledges. The other donors who had not sent money were willing to do so if the total $1 million was assured of collection. But no subscribers were available to replace Electric Lighting and Rosenwald. By 1932 it was certain that these pledges would not be honored. The fund was liquidated. In 1934 the few contributions paid in were returned to their donors. The campaign for the National Research Fund had ended.[46]

How is the demise of the National Research Fund to be interpreted? Helen Wright, who examined primarily Hale's role in the campaign, thinks "the plan fell victim to the times [that is, we presume, the depression] as well as to the lack of vision of the majority of industrial leaders." A. Hunter Dupree points, not so differently, to failures in leadership.[47] But in these terms, the fund did not fail. Although requiring four years, it succeeded in

Papers. Frank B. Jewett to Millikan, April 24, 1934, with enclosure, "Report to the Trustees of the National Research Fund and to the National Academy of Sciences, April 19, 1934, by F. B. Jewett," ibid.

45. Kellogg to Millikan, January 20, 1928, ibid.; "Minutes of the Meeting of the Trustees of the National Research Fund of the National Academy of Sciences, June 18th, 1928, at the University Club, New York, Paul Brackett, Recording Secretary," box 63, Hale Papers.

46. Jewett, "Report to the Trustees."

47. Wright, *Explorer of the Universe*, p. 369. Dupree, *Science in the Federal Government*, pp. 342–43.

obtaining pledges of a million dollars a year. If Electric Lighting and Rosenwald had met their pledges, the fund would have gone into operation, presumably, a year after the depression began.

The failure of the fund was ideological. The original goal of the campaign was to cultivate public awareness of the values and methods of the pure scientists. This goal was part of the general effort of the national scientists to convince the public of their ideology and thereby to promote cultural unity. The scientists thought that public recognition would bring financial support from the corporations. But the scientists never were able to establish a connection between public acceptance of their ideology and the corporate donations. Consequently, as the campaign for funds progressed from 1926 to 1929, the scientists' attention shifted away from constructing a primary relationship between themselves and the public to constructing such a relationship with the corporations. By 1929, their central concern was no longer to relate the method of pure science to the progressive values of individualism and democracy, but to demonstrate to industrialists that pure science, rather than engineering or applied science, was the basis of industrial profits.

The Limitations of Industrial Capitalism

One obstacle to obtaining public recognition of national science was that the social values to which the scientists appealed were preempted by the engineer. Moreover, the economic values to which pure scientists appealed were preempted by the engineer and applied scientist working within the corporate structure in the industrial research laboratory. Thus, when the Naval Advisory Board on Inventions was established, engineers and inventors were represented on it and physicists were not. And after the war, Hale could confide his worry to Merriam that engineers had a visible public identity and influence in government represented by Hoover, and the basic scientists did not. One response to this situation was the thematic emphasis in the popularization of science of the dependence of the engineer on the research of the pure scientist and the primacy of scientific knowledge over engineering technique. Another response was the campaign for the National Research Endowment.

Robert Millikan's analysis of the relationship between pure science and engineering contained the basis of the scientists' argument to the corporations that industrial progress was based on pure science. He theorized about this relationship when he assumed the headship of the California Institute (1921). The officials of the Carnegie Corporation which was a benefactor of the new institute did not fully understand what the school was. Another technical school? Or a theoretical institute? Why were theoretical and research science mixed with engineering? Millikan explained that the failures of American engineering were usually due to inadequate education of the engineers. In all America's technical schools, engineers were trained but were not educated to think up new problems or to strike off in new directions of research for answers. This educational deficiency was understandable, considering that the purpose of American engineering schools was to produce as large a number of engineers as possible for the ordinary and numerous engineering tasks which were presented in a developing industrial nation. But such an approach was inadequate for the problems of the new American civilization. The military technology of the war, for example, had presented problems which engineers could not solve because past engineering techniques were irrelevant. Similarly, electrical, radio, and telephonic technologies demanded wholly new engineering. In both instances, the new problems were solved by men who were perhaps called "engineers" but who nevertheless had been educated on the doctorate level in pure science research, not in engineering. Thus Frank B. Jewett, who had been chief "engineer" of Western Electric, had received his doctorate in physics under Millikan. "The technical schools of the country have made a blunder during the past two decades in sacrificing fundamentals in the endeavor to so train men in the details of industrial processes that they are ready to be producers the moment they leave school."[48]

48. Millikan to Henry S. Pritchett, acting president of the Carnegie Corporation, October 31, 1921, with the enclosure "The Opportunity of the California Institute of Technology" (Address of January 6, 1920), box 18, Millikan Papers.

An education in research in pure science was primary because the demand for qualitatively new engineering techniques always arose out of problems in basic science, that is, problems concerning the quest for knowledge of the natural world. Engineering research to develop new techniques obviously came after the presentation of the problems. For this reason, no engineer could do engineering research unless he could first do scientific research. "Two prominent mining engineers told me a few months ago," Millikan wrote to the Carnegie Corporation, "that if they had their engineering courses to take over again they would devote their time wholly to mathematics, physics, chemistry, biology, and language, and leave out all practical courses."[49] The guiding ideal of the California Institute of Technology was the education of a few well selected men in basic scientific research first, then in engineering. In this ideal, the institute was unique in America. The institute had another educational function as well. Millikan discussed at length what he considered wrong with the German graduate educational system. It was oriented around individual though perhaps great men, rather than around cooperative research. The ideal of the German system, Millikan apparently said, in a confused passage, was to produce men who viewed themselves as students of particular scientists rather than as members of a professional class. It was the latter which Millikan hoped the example of the California Institute would create in America.[50]

Millikan's rationalization of the priority of basic science over engineering and applied science clearly underlay the inclusion of the engineering division in the National Research Council. It was the basis for the argument to the industrialists that pure science was necessary for the progress of technology. If the engineering division were abolished or if pure science were not supported sufficiently, hope would be diminished for industrial development in America.[51]

The desire to educate the American people and the corporate leaders to the identity of the true benefactors of progress was a

49. Millikan to Pritchett, ibid.
50. "The Opportunity of the California Institute of Technology," ibid.
51. Millikan to Kellogg, November 27, 1922, box 125, Merriam Papers.

motivation as well as a goal of the campaign for the National Research Fund, appearing prominently in the campaign literature. The establishment of an operating fund would do more than simply provide money for research in pure science. It would provide a publicly legitimated institution identified with the pure scientists to which they could appeal for grants in the name of pure science, rather than in the name of engineering, applied science, or industrial research. Public confusion over the character of pure science due to the Einstein controversy would be alleviated to the extent that the pure scientists could be recognized by their association with the fund—an institution dispensing grants to a profession whose values had been related to other cultural values.

Considering George E. Hale's concern in 1920 and 1922 over the public awareness of the pure scientist, it is not surprising that when the National Academy in April 1925 authorized a study of the research fund proposal, the first action taken by Hale was to call a conference on the education of the public. John J. Carty could not attend, but he wrote encouragement to Hale: "In the minds of the American public, generally, proud of their practical achievements, the criterion of practical utility is the one applied to all scientific researches. There is need for education in this respect, for unless research in pure science is encouraged and pushed forward, the researches in applied science will soon suffer." Public allegiance to practicality meant that "any movement . . . calculated to interest the public generally in our scientific activities" had to be addressed specifically to this problem.[52]

The report of the National Academy committee on additional funds for research of November 1925 stated that the "chief object [of the proposed endowment] is to increase and strengthen American contributions to science by the creation of a national fund to aid skilled investigators engaged in fundamental research for the advancement of knowledge." But when this objective was stated, the scientists had to face several problems: "It is . . .

52. Carty to Hale, May 6, 1925, box 10, Hale Papers. The educational aspects of the National Research Fund (at this point still called the "endowment") were alluded to in Gano Dunn to Hale, May 6, 1925, box 14, Hale Papers.

evident that the accomplishment of our object involves the requirement of an intimate understanding of the present fundamental problems of science and a still more intimate appreciation of the abilities, characteristics, and needs of the investigators engaged in their solution." The campaign for the endowment was to be an exercise in understanding. The theories of relativity had revolutionized physics, and the public needed to know what "science" was. Einstein's theories, it was pointed out, were developed by the "most abstruse" mathematics. The American public, just emerging in 1925 from the strenuous controversy over relativity theory, and industrialists had to understand that such abstruse developments were the ultimate source of the "progress of civilization."[53]

What the scientists said in their campaign for the endowment corresponded closely to the themes of the popular science. The official announcement began with quotations by Elihu Root and Herbert Hoover reiterating one major theme: "Every practical advantage gained in utilizing natural forces for the benefit of mankind can be traced back to a necessary basis established through fundamental research in pure science by men who had no other object than to ascertain the truth;" "Applied science will dry up unless we maintain the sources of pure science." As one heading in the announcement stressed: "Pure Science the Source of Progress." The operation of the railroads, for example, was made possible by the "systematic observations of the stars for the determination of time." The pulp paper industry was based on "Réaumur's studies of wasps, which construct their paper nests of materials produced by the mastication of bits of wood." The announcement added what was thought to be a conclusive example: "Remove Galileo's discovery that force is measured by the product of mass by acceleration and the whole of modern civilization collapses like a house of cards, because not a steam engine, dynamo, or other dynamical device can be designed without it."[54]

53. Hale et al., "National Academy of Sciences: Report of the Committee on Additional Funds for Research," third and final draft, November 4, 1925, box 63, Hale Papers.
54. "National Research Endowment: A National Fund for the Support of Research in

The failure of the scientists to convince the corporate leaders of their ideology of science lay not simply in the historical and logical ambiguity of their examples of industry's indebtedness to pure science. The scientists' financial claims were never established in terms of the industrial, capitalistic framework that guided their would-be patrons' decisions. But the financial claims of the examples were yet more tenuous. The scientists were arguing that over the long run a certain financial investment would yield a greater profit and progress if given to pure science than if placed in applied science, industrial research, or engineering. The scientists considered this a valid claim because logically and methodologically all problems of applied research arose from pure scientific research. In this way, they thought they had destroyed the independence of applied research and the notion that the applied researcher or engineer was of greater technological and economic importance than the basic researcher. But the industrial leaders maintained a distinction between pure and applied sciences to which the pure scientists' examples were irrelevant. For the industrialists, applied science was research devoted to the development of commercial products. Pure science was research which did not have this goal. For the industrialist, the logical distinction between pure and applied sciences was less important than the commercial distinction. The scientists' argument that in the long run pure science yielded some practical product did not take into account the capitalistic criteria of development costs, sales, and profits. Industrial research decisions, including charity to research endowments, had to be made according to these criteria. If the ultimate objective was unspecified, no decision could be made within the framework of industrial capitalism. John J. Carty's argument that "pure scientific research is conducted with a philosophical purpose, for the discovery of truth, and for the advancement of learning"[55] would not convince a corporation to divert part of its research

---

Pure Science," n.d., box 193, Root Papers. A paraphrase and summary of the announcement appeared in "Say Pure Science Lags in America," *New York Times*, February 1, 1926, p. 7.
    55. Ibid.

budget from its own laboratory to a fund which would distribute grants to scientists making investigations unrelated to the corporation's products.

The rigid requirements imposed on the argument of the pure scientists by industrial capitalism were revealed as they modified their case for the corporations. To help Elihu Root overcome Elbert Gary's scruples, Millikan forwarded a short paper by John Johnston of Yale. It is not certain that Root used this paper on Gary, but if he did, almost certainly it weakened the scientists' position. The opening sentence went: "In the case of an ancient art such as the steel industry is, it is not easy to give a precise statement of the debt which it owes to science; though on the other hand, there can be no question that without the aid of science the condition and the magnitude of the industry would have been very different." With this disclaimer for a beginning, the remainder of the paper could not be effective. Johnston made references to alloys and to steel for high-speed tools, but these did not offset the introduction. No wonder that in September 1926, John J. Carty called for a more effective formulation of the scientists' argument for U.S. Steel.[56]

Millikan's argument for the indebtedness of the railroads to pure science was only a little better. "First, the steam engine itself became possible about the beginning of the nineteenth century," Millikan wrote, "only because a group of pure scientists, Galileo, Newton, Laplace, and others, through two centuries of effort (1600–1800) had worked out the basis of dynamics." He went on to mention that without the telegraph and astronomical timetables the railroads could not exist. Judged by capitalistic criteria this was a bad argument; the railroads never contributed to the research fund. At best, it epitomized the scientists' case, embellished by Millikan with metaphorical allusion to the dependence of progress on the frontier: "The pure scientist is merely the scout who is dispatched ahead for the sake of blazing out the trail along which a little later the army of engineers is to follow.

56. John Johnston, "The Indebtedness of the Steel Industry to Pure Science," enclosed in Millikan to Root, April 9, 1926, box 141, Root Papers; also in F. B. Jewett to Hoover, April 9, 1926, box 8, Millikan Papers. See note 28 above.

If America calls in her scouts her progress will cease and the demand for any increase in the services of the railroads will disappear."[57]

The scientists pushed their appeals into all business interests which might have the funds to contribute. Millikan made the greatest use of the examples of electronics industries because these were most unquestionably derived from fundamental investigations. The study of electron behavior in high vacua, in which Millikan himself had done his prizewinning research, led to the electron-vacuum tube and thus to the communications industry.[58] John J. Carty thought the insurance companies would be brought into the fund with the appeal to their self-interest, that improved health and social hygiene "are of direct advantage to [their] business."[59] But the insurance companies did not contribute.[60]

Most of the scientists were aware of the ineffectiveness of their appeal. As Carty wrote to Hale, "The immediate practical results from their researches [in pure science] are usually nil." It was also realized that the question of patenting stood in the way of industrial subscriptions to the fund. If a major discovery of immediate commercial value were made, who would have the patent to it? Were certain contributors to the fund to receive preferential treatment in access to the discovery?[61] Why should an industrialist contribute to a fund if discoveries financed by it might be to his competitor's advantage? Discoveries made in his own laboratory would at least be under his control. J. J. Carty fi-

---

57. Millikan, "Indebtedness of the Railroads to Pure Science," n.d., box 8, Millikan Papers. The occasion of this paper is clarified in Hoover to Millikan, July 2, 1926, and July 26, 1926, ibid.

58. Millikan, "Address Delivered over the Radio in Support of the Campaign of the University of Chicago, Dictated March 21, 1925"; and "The Dependence of the Progress of the World upon Pure Science Research, as Illustrated by the Recent Developments in the Art of Communication" (apparently April 1926), box 48, Millikan Papers.

59. Carty to Millikan, September 28, 1926, box 10, Millikan Papers.

60. The argument for basic research inside corporations was given a thorough treatment in Edward R. Weidlein and William A. Hamor, Science in Action: A Sketch of the Value of Scientific Research in American Industries (New York: McGraw-Hill Book Co., 1931). This book was not, however, arguing for national pure science as were the campaigners for the research fund.

61. Carty to Hale, September 29, 1926, box 10, Hale Papers. W. R. Whitney to Hoover, February 18, 1926, box 63, Hale Papers.

nally suggested that the scientists' case for the fund would be stronger if arrangements were made for money to be channeled into research most likely to lead to discoveries of benefit to its contributors.[62] Movement toward this arrangement had begun when Root, under pressure from Gary in the spring of 1926, secured a modification of the declaration of trust of the fund. According to this modification, subscribers to the fund of at least a million dollars could have representatives on an advisory committee to the fund's trustees with the right to make recommendations.[63] These alterations would drastically alter the character of the fund. The fund ceased to have as its objective the support of worthwhile pure science, regardless of topic. Instead, the modifications brought national science into the service of huge corporations as applied science (by corporate definition). Instead of securing the progress of all the people, the fund would secure only the progress of its contributors who benefitted by arrangements with it. Indeed, the final irony was that rather than educating the American people and industrialists to the values and requirements of pure science, the national scientists were themselves educated to the values and requirements of industrial capitalism. Whether the scientists would have been subservient to the wishes of the corporate contributors if the fund had actually gone into operation is a moot question; however, the trend of the campaign was not encouraging for their independence.

Consensus and Conservatism

Not only the limitations of industrial capitalism, but the limited range of American social and cultural values pushed the national scientists in a conservative direction. Louis Hartz has shown that political consensus on liberalism produced a limited range of symbolic values like "democracy" and "individualism." This consensus meant that there were no symbols or values other

---

62. Carty to Hale, September 29, 1926, box 10, Hale Papers.
63. Carty, Millikan, Dunn, Hale, Noyes, and Jewett, "Suggested Modification of the Declaration of Trust," in *Documents Submitted.*

than those of liberalism to which the scientists might legitimately appeal for support of their own scientific values. Despite the rhetoric of aristocracy in the post–1800 antebellum South, for example, that society never produced the scientific culture whose patronage had glorified European aristocracies. In part, of course, this was because the South never had the wealth, the urban culture, or the freedom necessary to support a class of scientists. Tocqueville's suggestion that the absence of an aristocracy in America was to blame for a lack of nonutilitarian science was accurate in the sense that there were no aristocratic values to which theoretical scientists could appeal to give their profession social and cultural importance if liberalism failed to do so.[64]

This characteristic of American democracy effectively controlled the potential of science for creating new symbols. The social history of science in Europe contains many accounts of a scientific value, such as "scientific method" or "scientific reason," becoming symbolic of revolutionary changes in society. For the eighteenth-century philosophe, Francis Bacon symbolized the revolt of experimental science against traditional authority. For Ludwig Buchner in the nineteenth century, materialistic science was symbolic of resistance to the forces of religious and political orthodoxy that crushed the revolution of 1848. This is not to deny that science can create symbols for conservatism, as reference to the "American School" of anthropology and to Pavlovian behaviorism indicates, but shows only that science does have the capacity to create revolutionary symbols.

At a time when progressivism was becoming conservative, the effort of the national scientists to relate scientific values to the values of American progressivism had two consequences: the potential for creating new symbols was effectively suppressed, and scientific values themselves appeared conservative. Instead of creating a new symbol of cultural unity of science and nonscience, the scientists attempted to take Herbert Hoover away from

64. Thomas C. Johnson, *Scientific Interests in the Old South* (New York: D. Appleton-Century, 1936); Clement Eaton, *The Freedom-of-Thought Struggle in the Old South* (New York: Harper & Row, Torchbooks, 1964), pp. 304–07; Alexis de Tocqueville, *Democracy in America*, ch. 22.

the engineers to be such a symbol. His activities and reorganization of the Department of Commerce during two administrations symbolized the engineer's ability to "get things done." When he ran for president and could devote less time to the endowment campaign, the scientists' opportunity to capture him as a symbol diminished. So Bertrand Snell, Republican caucus leader in the House of Representatives, hailed Hoover as "the engineer President of the United States, [who] is solving, and will solve, stupendous and vexatious problems."[65] As Hoover and the engineer became symbols of conservatism, the scientists' effort to have Hoover represent science appeared conservative.

Another instance of symbol competition was the attempt of the national scientists through the war and the 1920s to take the scientific method away from the engineers, social scientists, and management experts who had possessed it in the prewar progressive period. In progressive thought, the "scientific method" was a key symbol holding together disparate values, as Robert Wiebe in *The Search for Order* has argued.[66] Slosson, in his popularizations, and Millikan and Merriam, in their discussions of scientific method, emphasized the dependence of engineering and applied science on basic science. They thought progress was dependent on the method of pure science, not the technique of technology. The campaign for the research endowment was a final effort to convince the public and the business leaders that pure scientists were the true possessors of the scientific method.

Scientific values themselves appeared conservative when the scientists were unable to link their relationship with the lay public to their relationship with the corporations. In the progressive period, the social scientist, engineer, and the expert had all provided means by which the people could control the economic and social position of big business. In the postwar decade, the national scientists were unable to provide any means for this

65. Quoted in Charles A. Beard and Mary R. Beard, *America in Midpassage* (New York: Macmillan, 1939), p. 125. The Beards have a good discussion of Hoover the engineer, pp. 95-96, 121 ff. My italics in text.

66. Robert H. Wiebe, *The Search for Order, 1877-1920* (New York: Hill and Wang, 1967).

control. The scientists did not convince the corporations to support an independent profession of pure science which would guarantee the general welfare. Rather than being the instruments of the welfare of the people, the pure scientists became the instruments of the welfare of the corporation.

The failure of the scientists to build a consensus for progressive democracy based on national science was indicated in the 1930s by the relative oblivion into which national science sank. The public stature of Robert Millikan declined precipitously. Charles A. Beard turned to Millikan in 1930 for a statement defending the benefits of science, as if these benefits could be doubted. But by 1939, few people would even listen to Millikan. In 1929, Beard had edited a collection of papers whose theme was anxiety over the future of a scientific technological civilization and whose title was *Whither Mankind?* The response to the title question came in 1930 from engineers, not scientists, who were considered the champions of American progress: *Toward Civilization.*[67]

Anxieties over American civilization created by the depression were not concerned with pure science, but with technology. The technocracy movement ignored basic science and looked to rational engineering to bring America out of the depression.[68] This indicates again how unsuccessful the national scientists had been in convincing the American people that science was the basic cause of cultural change.

The New Deal reinforced this tendency to ignore basic science. The physical sciences on which the national scientists based their ideology were placed in the background. Efforts to gain Public Works Administration assistance for physical science were largely unsuccessful. Secretary of Agriculture Henry A. Wallace, who systematically developed a conception of the place of science in

67. Charles A. Beard, ed., *Toward Civilization* (New York: Longmans, Green, 1930), "Science Lights the Torch," by Robert A. Millikan, pp. 38–46. Beard and Beard, *America in Midpassage*, p. 853. Charles A. Beard, ed., *Whither Mankind: A Panorama of Modern Civilization* (1928; reprint ed., New York: Longmans, Green, 1937).

68. Henry Elsner, Jr., *The Technocrats: Prophets of Automation* (Syracuse: Syracuse University Press, 1967), covers the organizational aspects of the movement, but is unsatisfactory on the ideology.

government, conceived of science primarily in terms of social science and social engineering. Attempts of scientists on the president's Science Advisory Board (created in 1933 and largely financed by the Rockefeller Foundation) failed to make nongovernmental scientists relevant to the emergency of the depression. The separation of the physical scientists from American culture and their irrelevance to the 1930s was illustrated not only by the opposition of scientists to the New Deal, but by the difficulty they had in conveying to the president their fears about a German atomic bomb.[69]

Consequences of the Scientists' Failure

By 1930 it was apparent that the scientists had failed to create a public consensus on the ideology of national science. Under the direction of Edwin Slosson, Science Service popularized the values and method of pure science. But from the beginning, this effort was defeated by the controversy over Einstein's theories of relativity. The effort of Robert Millikan to anchor the basic American value of progress in the cosmos, and in the scientific method which revealed the cosmos, had likewise been undercut by the revolution begun by the theories of relativity. Paul R. Heyl's loss of faith in the scientists' ideology testified to the intellectual tensions within it. The national scientists endeavored to prove that science reinforced the basic social and economic values of American progressive democracy. The weakness in the relationship was exposed when his loss of faith in democracy brought Robert Millikan to believe in scientific oligarchy. Finally, the ordeal of soliciting money from the big corporations in the endowment campaign pushed the scientists to subvert their own independence and their primary relationship to the general wel-

69. Dupree, *Science in the Federal Government*, pp. 344–68; Lewis E. Auerbach, "Science in the New Deal: A Pre-War Episode in the Relations Between Science and Government in the United States," *Minerva* 3 (Summer 1965): 457–82. On the political opinions of Lawrence and other scientists as well as a recent survey of the well-known efforts of the scientists to reach Roosevelt in 1939, see Nuel Pharr Davis, *Lawrence and Oppenheimer* (New York: Simon and Schuster, 1968).

fare. They were unable to demonstrate precisely how direct public support of basic research would preserve political democracy from corruption by extra-political interest or how basic research would increase the economic prosperity of the average American. Instead, they were able only vaguely to relate their services to the long-term interest of gigantic corporations which were hardly concerned with the general welfare.

The failure of the national scientists' ideological mission in the 1920s had three important consequences for American culture. The values and method of pure science were not integrated into the consensus of liberalism. Engineering and technocracy continue to possess the name of science. And, the conclusion that war (hot or cold) is the only justification for national science became inevitable.

The exclusion of the values and method of pure science from American liberalism is revealed in the contemporary anxiety over the split between the two cultures. It is not necessary to list popular and intellectual literature to make this point. A single quotation will be ample. Gerald Holton, professor of physics at Harvard, complains that "loss of cohesion is perhaps the most relevant symptom of the disease of our culture; . . . this is a sort of failure of image."[70] The scientists' emotions of atomization and isolation were of course central in their response to the exhilarating experience of cultural unity in the First World War. Their cultural situation had been dissected brilliantly a few years earlier by Herbert Croly in *The Promise of American Life*. The contemporary isolation of the basic scientists is not primarily the result of anguish over the atomic bomb. It is the result of the decline of natural theology and popularization while science was being professionalized in the last century and the technocratization of the scientific method in this century.

Gerald Holton and others are partially correct when they see this separation of cultures as due to a "failure of image."[71] But

70. Gerald Holton, "Modern Science and the Intellectual Tradition," in *The New Scientist: Essays on the Methods and Values of Modern Science*, ed. Paul C. Obler and Herman A. Estrin (Garden City, N.Y.: Doubleday, Anchor Books, 1962), p. 23.
71. Ibid., pp. 23–33. Incorrect public images of science and the scientists are also dis-

this failure is not simply the public acceptance of an inaccurate vision of the personal character, values, and method of the scientist. This sort of ignorance had bothered scientists before the First World War and they sought to remedy it by the popularization in the 1920s. The bad images of the scientist today—the iconoclast, the ethical pervert[72]—will persist as long as science occasionally contributes to man's self-destruction. Rather, the "failure of image" must be understood in a more profound sense. The failure is the inability to possess those symbols and values which can unite the two cultures. This is, and was in the 1920s, an ideological problem, having to do with the definition and consistency of values, their relationship to reality and to society, and man's perception of reality; it is not a problem of public relations and advertising. American liberalism has considered itself scientific since the progressive era, but this scientific character has sprung from social science, scientific management, and rational engineering, not from the physical sciences with which the scientific method originated. The struggle of the scientists in the 1920s to regain possession of the symbols of science and the scientific method represented their awareness that the values and method of pure science had to have a naturally sanctioned connection to the social values of democracy if the two cultures were to have a deep unity. Their failure was due to the new physics which destroyed the classical vision of revelatory science, to the conservative pressures of a disintegrating progressivism, and to industrial capitalism in the 1920s. The separation of the two cultures can never be overcome if scientists think in terms of public relations or of education to values which the struggle of the 1920s showed not to be within the democratic consensus.

The name of science remains primarily in the possession of engineering and technocracy. An indication of this exists in governmental aid to science. Recent works have carefully explored the matter of governmental aid. Pure science, as Daniel Greenberg says, is constantly striving to "redefine itself and to distin-

cussed in Dael Wolfe, "Science and the Public Understanding," in *The New Scientist*, pp. 117–26.

72. Holton, "Modern Science and the Intellectual Tradition," pp. 25–26.

guish its essential qualities from those of other technical activities."[73] That an intellectual activity over three hundred years old should now have an "identity crisis" reveals the depth of pure science's ideological problem in American democracy. America has not provided the symbols, tasks, or institutions which could establish the place of pure science within the culture. Consequently, pure science must constantly redefine itself both to be distinguished and to be eligible for the money which government grants to certain scientific activities.

Daniel Greenberg and H. L. Nieburg both discuss the final consequence of the scientists' failure in the 1920s: that war, hot or cold, is the only justification for national science.[74] War had united the two cultures from 1915 to 1918. Military competition created the many technological problems, like submarine detection, which demanded fundamental scientific research on a national scale. War united pure scientific research with national goals. The impact of the First World War on the scientists was a preview of the similar war experience of the scientists twenty-five years later. If the appeal to the war experience failed to achieve continued unification of the two cultures and support for national science after World War I, it was only because America returned to normalcy. The argument that international competition demanded national science was the only argument not undermined by developments within science, the symbolic capture of "science" by engineers and technocrats, or the disintegration of progressive democracy.

When World War II began, the new generation of scientists, who had been doing graduate study in the 1920s, would undergo an experience similar to that of the previous generation. They would emerge, however, into a cold war which maintained their close unity with the other sectors of American society, through crises of thermonuclear weapons and sputniks. But when the cold war entered a thaw in the 1960s, the justification of national science by international competition would be less pressing, and the old two-cultures division would reappear.

73. Daniel S. Greenberg, *The Politics of Pure Science* (New York: New American Library, 1967), p. 10.
74. H. L. Nieburg, *In the Name of Science* (Chicago: Quadrangle Books, 1966).

SELECTED BIBLIOGRAPHY

INDEX

SELECTED BIBLIOGRAPHY

Manuscript Sources

California Institute of Technology Archives
    Robert Andrews Millikan Papers
Cornell University Archives
    Ernest Merritt Collection
Henry E. Huntington Library and Art Gallery
    George Ellery Hale Collection
Library of Congress, Manuscript Division
    James McKeen Cattell Papers
    Scudder Klyce Papers
    Jacques Loeb Papers
    John Campbell Merriam Papers
    Simon Newcomb Papers
    Elihu Root Papers
Mount Wilson and Palomar Observatories Archives and the California Institute of Technology Archives
    George Ellery Hale Papers (Available also on microfilm. *The George Ellery Hale Papers, 1882–1937*, edited by Daniel J. Kevles. Washington, D.C. and Pasadena, Calif.: Carnegie Institution of Washington and the California Institute of Technology, 1968.)

Magazines

Atlantic Monthly
Century
Collier's
Current Opinion
Forum
Harper's Monthly Magazine
Independent
Literary Digest
Nation

New Republic
Outlook
Popular Science Monthly (1872–1915; in 1915 this journal became the *Scientific Monthly*)
Saturday Review of Literature
Science
Science News Letter
Scientific American
Scientific Monthly
Scribner's Magazine
Vital Speeches of the Day
World's Work

Primary Published Sources

BOOKS

Abbott, Edwin Abbott. *Flatland: A Romance of Many Dimensions.* Date of first edition not established; 2d rev. ed., 1885. Reprint. New York: Dover Publications, 1964.

Adams, Henry. *The Degradation of the Democratic Dogma.* Edited with an introduction by Brooks Adams. New York: Macmillan, 1919.

———. *The Education of Henry Adams: An Autobiography.* Boston: Houghton Mifflin, 1927.

Beard, Charles A., ed. *Toward Civilization.* New York: Longmans, Green, 1930.

———. *Whither Mankind: A Panorama of Modern Civilization.* 1928. Reprint. New York: Longmans, Green, 1937.

Beard, Charles A. and Mary R. *America in Midpassage.* New York: Macmillan, 1939.

Bird, James Malcolm, ed. *Einstein's Theories of Relativity and Gravitation: A Selection of Material from the Essays Submitted for the Eugene Higgins Prize of $5,000.* New York: Scientific American Publishing Co., 1922.

Bridgman, P. W. *The Logic of Modern Physics.* 1927. Reprint. New York: Macmillan, 1961.

———. *The Nature of Thermodynamics.* New York: Harper, 1961.

Caldwell, Otis William, and Slosson, E. E., eds. *Science Remaking the World.* Garden City, N.Y.: Garden City Publishing Co., 1923.

Carmichael, Robert D.; Davis, Harold T.; MacMillan, William D.; and Hufford, Mason E. *A Debate on the Theory of Relativity.* Chicago: Open Court Publishing Co., 1927.

Cohen, Morris Raphael. *Reason and Nature: An Essay on the Meaning of Scientific Method.* 1931. Reprint. London: Collier-Macmillan, 1964.

———. *Studies in Philosophy and Science.* New York: Henry Holt, 1949.

Croly, Herbert. *The Promise of American Life.* 1909. Reprint. New York: E. P. Dutton, 1963.

Dewey, John. *Essays in Experimental Logic.* Chicago: University of Chicago Press, 1916.

Dingle, Herbert. *Relativity for All.* London: Methuen, 1922.

Duncan, Robert Kennedy. *The New Knowledge: A Popular Account of the New Physics and the New Chemistry in Their Relation to the New Theory of Matter.* New York: A. S. Barnes, 1905.

Eddington, Arthur Stanley. *Space, Time and Gravitation: An Outline of the General Relativity Theory.* 1920. Reprint. New York: Harper & Row, 1959.

Einstein, Albert. *Essays in Science.* New York: Wisdom Library, 1934.

Fosdick, Raymond Blaine. *The Old Savage in the New Civilization.* Garden City, N.Y.: Doubleday, Doran, 1929.

Heyl, Paul R. *The Common Sense of the Theory of Relativity.* Baltimore: Williams & Wilkins, 1924.

———. *The Fundamental Concepts of Physics in the Light of Modern Discovery.* Three Lectures at the Carnegie Institute of Technology, Pittsburgh, January 1925. Baltimore: Williams & Wilkins, 1926.

———. *The New Frontiers of Physics.* New York: D. Appleton, 1930.

———. *The Philosophy of a Scientific Man.* New York: Vanguard Press, 1933.

Hoover, Herbert Clarke. *American Individualism.* Garden City, N.Y.: Doubleday, Page, 1922.

Hopkins, N. M. *The Outlook for Research and Inventions.* New York: Van Nostrand, 1919.

Jeans, James Hopwood. *Astronomy and Cosmogony.* 1928. Reprint. Cambridge, England: University Press, 1929.

———. *The Mysterious Universe.* New York: Macmillan, 1930.

Knight, Oliver, ed. *I Protest: Selected Disquisitions of E. W. Scripps.* Madison: University of Wisconsin Press, 1966.

Lippmann, Walter. *Drift and Mastery.* 1914. Reprint. Englewood, N.J.: Prentice-Hall, 1961.

———. *Public Opinion.* 1922. Reprint. New York: Macmillan, 1961.

Lorentz, H. A.; Einstein, A.; Minkowski, H.; and Weyl, H. *The Principle of Relativity: A Collection of Original Memoirs on the Special and General Theory of Relativity.* Translated by W. Perrett and G. B. Jeffrey. 1923. Reprint. New York: Dover Publications, n.d.

McCarthy, Charles. *The Wisconsin Idea.* New York: Macmillan, 1912.

McConnell, Francis John; Woodbridge, Frederick J. E.; Pound, Roscoe; Taft, Lorado; Millikan, Robert A.; and Shorey, Paul E. *The Creative Intelligence and Modern Life.* Boulder: University of Colorado Press, 1928.

Magie, W. F. *Principles of Physics.* New York: Century, 1911.

———. ed. *The Second Law of Thermodynamics: Memoirs by Carnot, Clausius, and Thomson.* New York: Harper & Brothers, 1899.

Manning, Henry Parker, ed. *The Fourth Dimension Simply Explained: A Collection of Essays Selected from Those Submitted in the Scientific American's Prize Competition.* 1910. Reprint. New York: Dover Publications, 1960.

Merriam, John Campbell. *Published Papers and Addresses.* 4 vols. Washington, D.C.: The Carnegie Institution of Washington, 1938.

Millikan, Robert Andrews. *Autobiography.* New York: Prentice-Hall, 1950.

———. *Cosmic Rays: Three Lectures, Being the Revision of the 1936 Page-Barbour Lectures of the University of Virginia and the 1937 John Joly Lectures of Trinity College, Dublin.* Cambridge, England: University Press, 1939.

———. *The Electron, Its Isolation and Measurement and the Determination of Some of Its Properties.* Chicago: University of Chicago Press, 1917.

———. *Electrons (+ and −), Protons, Photons, Neutrons, Mesotrons, and Cosmic Rays.* 1935. Reprint. Chicago: University of Chicago Press, 1947.

————. *Evolution in Science and Religion.* New Haven: Yale University Press, 1927.

————. *Science and Life.* Boston: Pilgrim Press, 1924.

————. *Science and the New Civilization.* New York: Charles Scribner's Sons, 1930.

————. *A Scientist Confesses His Faith.* Science and Religion Series of Popular Religion Leaflets. Chicago: American Institute of Sacred Literature, 1923.

————. *Time, Matter and Values.* Chapel Hill: University of North Carolina Press, 1932.

Mumford, Lewis. *Technics and Civilization.* 1934. Reprint. New York: Harcourt, Brace & World, Harbinger Books, 1963.

Northrup, F. S. C., and Gross, Mason W., eds. *Alfred North Whitehead: An Anthology.* New York: Macmillan, 1961.

Poffenberger, A. T., ed. *James McKeen Cattell, 1860–1944: Man of Science.* 2 vols. Lancaster, Pa.: Science Press, 1947.

Poor, Charles Lane. *Gravity Versus Relativity: A Nontechnical Explanation of the Fundamental Principles of Gravitational Astronomy and a Critical Examination of the Generalized Theory of Relativity.* New York: G. P. Putnam's Sons, Knickerbocker Press, 1922.

Pupin, Michael. *From Immigrant to Inventor.* New York: Charles Scribner's Sons, 1923.

————. *The New Reformation: From Physical to Spiritual Realities.* New York: Charles Scribner's Sons, 1927.

————. *Romance of the Machine.* New York: Charles Scribner's Sons, 1930.

Ratner, Joseph, ed. *Intelligence in the Modern World: John Dewey's Philosophy.* New York: Modern Library, 1939.

Reichenbach, Hans. *The Direction of Time.* Edited by Marie Reichenbach. Berkeley: University of California Press, 1956.

Robinson, James Harvey. *The Mind in the Making: The Relation of Intelligence to Social Reform.* New York: Harper & Brothers, 1921.

————. *The New History: Essays Illustrating the Modern Historical Outlook.* New York: Macmillan, 1921.

Russell, Bertrand. *The A B C of Relativity.* New York: Harper & Brothers, 1925.

Slosson, Edwin E. *The American Spirit in Education: A Chronicle of Great Teachers.* New Haven: Yale University Press, 1921.

————. *Bibliography of Writings of Edwin Emery Slosson.* Washington, D.C.: Science Service, 1929.

————. *Chats on Science.* New York: Century, 1924.

————. *Creative Chemistry: Descriptive of Recent Achievements in the Chemical Industries.* New York: Century, 1919.

————. *Easy Lessons in Einstein: A Discussion of the More Intelligible Features of the Theory of Relativity.* New York: Harcourt, Brace and Howe, 1920.

————. *Great American Universities.* New York: Macmillan, 1910.

————. *Keeping Up with Science: Notes on Recent Progress in the Various Sciences for Unscientific Readers.* New York: Blue Ribbon Books, 1924.

————. *Major Prophets of Today.* Boston: Little, Brown, 1914.

————. *A Number of Things.* Arranged and with a biographical memoir by Preston William Slosson. New York: Harcourt, Brace, 1930.

————. *A Plea for Popular Science.* Washington, D.C.: Science Service, 1920.

————. *Plots and Personalities: A Method of Testing and Training the Creative Imagination.* New York: Century, 1922.

————. *Sermons of a Chemist.* New York: Harcourt, Brace, 1925.

————. *Short Talks on Science.* New York: Century, 1930.

————. *Snapshots of Science*. New York: Century, 1928.

Soddy, Frederick. *The Interpretation of Radium*. 1909. Reprint. London: John Murray, 1912.

————. *Science and Life: Aberdeen Addresses*. London: John Murray, 1920.

Taylor, Frederick W. *The Principles of Scientific Management*. New York: Harper & Brothers, 1911.

————. *Scientific Management*. New York: Harper & Brothers, 1947.

Tocqueville, Alexis de. *Democracy in America*. The Henry Reeves text as revised by Francis Bowen, now further corrected and edited with an introduction, editorial notes, and bibliographies by Phillips Bradley. 2 vols. New York: Alfred A. Knopf, 1945.

Tolman, Richard Chace. *Relativity, Thermodynamics and Cosmology*. Oxford: Clarendon Press, 1934.

Tuckerman, Alfred. *Index to the Literature of Thermodynamics*. Washington, D.C.: Smithsonian Institution, 1890.

Weidlein, Edward R., and Hamor, William A. *Science in Action: A Sketch of the Value of Scientific Research in American Industries*. New York: McGraw-Hill Book Co., 1931.

Weiner, Philip P., ed. *Values in a Universe of Chance: Selected Writings of Charles S. Peirce*. Garden City, N.Y.: Doubleday, Anchor Books, 1958.

Wells, H. G. *The Outline of History, Being a Plain History of Life and Mankind*. 2 vols. New York: Macmillan, 1920.

Yerkes, Robert M., ed. *The New World of Science: Its Development During the War*. New York: Century, 1920.

ARTICLES

Abbott, Ernest Hamlin. "The Fair at St. Louis." *Outlook* 74 (July 4, 1903): 552–63.

Able, Edmund Edward Fournier D'. *See* Fournier D'Able.

"An Alternative to Einstein: How Dr. Poor Would Save Newton's Law and the Classical Time and Space Concept." *Scientific American* 124 (June 11, 1921): 468.

"The Anti-Einstein Campaign." *Scientific American* 124 (May 14, 1921): 382.

Baird, J. W. " 'Popular' Science." *Science*, n.s. 26 (July 19, 1907): 75–76.

Berget, Alphonse. "The Principle of Relativity." *Scientific American Supplement* 83 (June 30, 1917): 411.

Birkhoff, George D. "The Origin, Nature and Influence of Relativity." *Scientific Monthly* 18 (March-June 1924): 225–38, 408–21, 517–28, 616–24; 19 (July-August 1924): 18–29, 180–87.

Bragdon, Claude. "New Concepts of Space and Time." *Dial* 68 (February 1920): 187–91.

"The Carnegie Institution." *Nation* 78 (January 14, 1904): 26.

Cattell, James McKeen. "Science, Education and Democracy." *Science*, n.s. 39 (January 30, 1914): 154–64.

[————.] "Scientific Journals and the Public." *Popular Science Monthly* 87 (September 1915): 309–10.

————. "Scientific Research and Sigma Xi." *Science*, n.s. 41 (May 14, 1915): 729–32.

"Changing the Mind Gears." *Literary Digest* 64 (January 24, 1920): 29.

Comstock, Daniel F. "The Principle of Relativity." *Science*, n.s. 31 (May 20, 1910): 767–72.

Cooke, Morris Llewellyn. "The Engineer and the People: A Plan for a Larger Measure of Cooperation Between the Society and the General Public." *Transactions, American Society of Mechanical Engineers* 30 (1908): 619–37.

D'Able, Edmund Fournier. *See* Fournier D'Able.

"Einstein's Finite Universe." *Scientific American* 124 (March 12, 1921): 202.

"Exit the Amateur Scientist." *Nation* 83 (August 23, 1906): 159–60.

Farlow, W. G. "The Popular Conception of the Scientific Man at the Present Day." *Science*, n.s. 23 (January 5, 1906): 1–14.

"Finding the Exceptional Man." *Independent* 54 (November 27, 1902): 2847–49.

Fournier D'Able, Edmund Edward. "The Principle of Relativity: A Revolution in the Fundamental Concepts of Physics." *Scientific American Supplement* 72 (November 11, 1911): 319.

Hale, George Ellery. "Cooperation in Research." *Science*, n.s. 51 (February 13, 1920): 149–55.

———. "How Men of Science Will Help in Our War." *Scribner's Magazine* 61 (June 1917): 721–26.

———. "Industrial Research and National Welfare." *Science*, n.s. 48 (November 22, 1918): 505–07.

———. "The International Organization of Research." In *The New World of Science: Its Development During the War*, edited by Robert M. Yerkes. New York: Century, 1920.

———. Introduction to *The New World of Science: Its Development During the War*, edited by Robert M. Yerkes. New York: Century, 1920.

———. "National Academies and the Progress of Research. I. The Work of the European Academies." *Science*, n.s. 38 (November 14, 1913): 681–98.

———. "National Academies and the Progress of Research. II. The First Half Century of the National Academy of Sciences." *Science*, n.s. 39 (February 6, 1914): 189–200.

———. "National Academies and the Progress of Research. III. The Future of the National Academy of Sciences." *Science*, n.s. 40 (December 25, 1914): 907–19.

———. "National Academies and the Progress of Research. II." *Science*, n.s. 41 (January 1, 1915): 12–23.

———. "National Academies and the Progress of Research: II." *Science*, n.s. 41 (January 1922): 515–31.

———. "The National Research Council." *Science*, n.s. 44 (August 25, 1916): 264–66.

———. "The Possibilities of Cooperation in Research." In *The New World of Science: Its Development During the War*, edited by Robert M. Yerkes. New York: Century, 1920.

———. "The Proceedings of the National Academy as a Medium of Publication." *Science*, n.s. 41 (June 4, 1915): 815–17.

———. "The Responsibilities of the Scientist." *Science*, n.s. 50 (August 15, 1919): 143–46.

———. "Science and the Wealth of Nations." *Harper's Magazine* 156 (January 1928): 243–51.

———. "Science and War." In *The New World of Science: Its Development During the War*, edited by Robert M. Yerkes. New York: Century, 1920.

———. "War Services of the National Research Council." In *The New World of Science: Its Development During the War*, edited by Robert M. Yerkes. New York: Century, 1920.

Hamor, William A. "The Value of Industrial Research." *Scientific Monthly* 1 (October 1915): 86–92.

Henderson, Archibald. "Is Einstein Wrong?—A Debate: II. The Triumphs of Relativity." *Forum* 72 (July 1924): 13–21.

Heyl, Paul R. "Cause or Chance?" *Scientific Monthly* 34 (March 1932): 273–77.

———. "Common Sense of the Theory of Relativity." *Scientific Monthly* 17 (December 1923): 513–26.

———. "Cosmic Emotion." *Scientific Monthly* 55 (December 1942): 558–65. Also in *Journal of the Washington Academy of Sciences* 32 (August 1942): 221–28.

———. "Fundamental Concepts in Physics in the Light of Recent Discoveries." *Science*, n.s. 61 (February 27, 1925): 221–25. [This article is an abstract of three lectures delivered at the Carnegie Institute of Technology, January 6, 7, and 8, 1925, and parallels the book by Heyl of the same title.]

———. "The Genealogical Tree of Modern Science." *Journal of the Washington Academy of Sciences* 33 (November 1943): 327–34.

———. "The Humanist and and I." *Scientific Monthly* 21 (August 1925): 173–76.

———. "Master Key." *Scientific Monthly* 19 (July 1924): 5–17.

———. "The Mystery of Evil." *Open Court* 34 (January–March 1920): 34–48, 74–86, 155–69.

———. "The Perspective of Modern Physics." *Scientific American* 145 (September 1931): 168–70.

———. "The Present Status of the Theory of Relativity." *Scientific Monthly* 23 (July 1926): 65–70.

———. "Report of the Committee on the 1929 Revision of the Academy's List of One Hundred Popular Books in Science." *Journal of the Washington Academy of Sciences* 19 (June 4, 1929): 207–17.

———. "The Riddle of Empty Space." *Science Digest* 7 (June 1940): 29–35. [Same article as "The Space in Which We Live."]

———. "Romance or Science?" *Journal of the Washington Academy of Sciences* 23 (February 1933): 73–83. Also in *Annual Report of the Board of Regents of the Smithsonian Institution*. Washington, D.C.: Government Printing Office, 1933, pp. 283–92.

———. "Skeptical Physicist." *Journal of the Washington Academy of Sciences* 28 (March 1938): 77–83. Also in *Scientific Monthly* 46 (March 1938): 225–29.

———. "The Solid Ground of Nature." *Scientific Monthly* 25 (July 1927): 25–33.

———. "The Space in Which We Live." *Scientific Monthly* 50 (March 1940): 251–57. [Same article as "The Riddle of Empty Space."]

———. "Space, Time and Einstein." *Scientific Monthly* 29 (September 1929): 230–35.

———. "The Strangest Thing in Physics." *Scientific American* 140 (June 1929): 498–500.

———. "The Student of Nature." *Scientific Monthly* 24 (June 1927): 497–506.

———. "Visions and Dreams of a Scientific Man." *Scientific Monthly* 26 (June 1928): 514–20.

———. "What Is Gravitation?" *Scientific Monthly* 47 (August 1938): 114–23.

———. "The Wonder of the Commonplace." *Scientific American* 135 (October 1926): 250–51.

Howe, Frederic C. "The World's Fair at St. Louis, 1904." *Cosmopolitan* 35 (July 1903): 277–90.

Humphreys, William J. "Right and Wrong in Popular Science Books." *Science*, n.s. 29 (February 19, 1909): 297.

———. "What Is the Principle of Relativity?" *Scientific American* 106 (June 8, 1912): 525–26.

"Is Einstein's Arithmetic Off?" *Literary Digest* 83 (November 8, 1924): 20–21.

"Is Einstein Wrong? A Symposium: Summarizing or quoting opinions of many scientists on a subject which was debated by Charles Lane Poor and Archibald Henderson in the June and July numbers of *The Forum*." *Forum* 72 (August 1924): 277–81.

Lee, Joseph. "Democracy and the Expert." *Atlantic Monthly* 102 (November 1908): 611–20.

Lotka, A. J. "A New Conception of the Universe." *Harper's Magazine* 140 (March 1920): 477–87.

McAdie, Alexander. "Relativity and the Absurdities of Alice." *Atlantic Monthly* 127 (June 1921): 811–14.

MacMillan, William Duncan. "Cosmic Evolution, First Part: What Is the Source of Stellar Energies?" *Scientia* (Milan) 33 (January 1923): 3–12.

———. "Cosmic Evolution, Second Part: The Organization and Dissipation of Matter Through the Agency of Radiant Energy." *Scientia* (Milan) 33 (February 1923): 103–12.

———. "The Fourth Doctrine of Science and Its Limitations." In *A Debate on the Theory of Relativity*, by Robert D. Carmichael, Harold T. Davis, William D. Mac-Millan, and Mason E. Hufford. Chicago: Open Court Publishing Co., 1927.

———. "The New Cosmology." *Scientific American* 134 (May 1926): 310–11.

———. "The Postulates of Normal Intuition: The First Speech of the Negative." In *A Debate on the Theory of Relativity*, by Robert D. Carmichael, Harold T. Davis, William D. MacMillan, and Mason E. Hufford. Chicago: Open Court Publishing Co., 1927.

———. "Some Mathematical Aspects of Cosmology, I. Cosmogony." *Science*, n.s. 62 (July 24, 1925): 63–72.

———. "Some Mathematical Aspects of Cosmology, II. Cosmology." *Science*, n.s. 62 (July 31, 1925): 96–99; 62 (August 7, 1925): 121–27.

———. "Stellar Evolution: An Attempt to Correlate Two Outstanding Problems of Physics." *Scientific American Supplement* 87 (May 24, 1919): 322.

———. "The Structure of the Universe." *Science*, n.s. 52 (July 23, 1920): 67–74.

Magie, William Francis. "The Primary Concepts of Physics." *Science*, n.s. 35 (February 23, 1912): 281–93.

Merriam, John Campbell. "Application of Science in Human Affairs." 1938. *Published Papers and Addresses* (cited hereafter as *PPA*) 4: 2124–34. Washington, D.C.: The Carnegie Institution of Washington, 1938.

———. "Bibliography of John Campbell Merriam." *PPA* 4: 2631–51.

———. "The Breadth of an Education." 1922. *PPA* 4: 2062–68.

———. "Charts and Compasses." 1933. *PPA* 4: 2078–85.

———. "Common Aims of Culture and Research in the University." 1922. *PPA* 4: 2386–95.

———. "The Evolution of Society as Influenced by the Engineer." 1933. *PPA* 4: 2409–15.

———. Foreword to *Science and the Public Mind*, by Benjamin C. Gruenberg. New York: McGraw-Hill Book Co., 1935. *PPA* 4: 2456–58.

———. "The Inquiring Mind in a Changing World." 1934. *PPA* 4: 2094–103.

———. "Making a Living—or Living?" 1930. *PPA* 4: 2032–38.

———. "Medicine and the Evolution of Society." 1926. *PPA* 4: 2396–408.

———. "The National Academy of Sciences—Dedication of Building." 1924. *PPA* 4: 2047–50.

———. "The Opportunities of the Federal Government in Research." 1930. *PPA* 4: 2469–75.

————. "The Place of Research in the Progress of the Next Generation." 1931. *PPA* 4: 2435–42.

————. "Reality in Adult Education." 1933. *PPA* 4: 2443–45.

————. "The Relation of Science to Technological Trends." 1937. *PPA* 4: 2490–93.

————. "Remarks at Science Service Round-Table Conference." 1932. *PPA* 4: 2045–46.

————. "Remarks by Chairman Introducing Alfred Noyes, Conference on the Obligation of Universities to the Social Order." 1933. *PPA* 4: 2061.

————. "The Research Spirit in Everyday Life of the Average Man." 1920. *PPA* 4: 2376–85.

————. "Science and Culture." 1934. *PPA* 4: 2086–93.

————. "Science and Government." 1933. *PPA* 4: 2476–79.

————. "Science and Human Values." 1936. *PPA* 4: 2116–23.

————. "Science and the Constructive Life." 1933. *PPA* 4: 2069–77.

————. "The Search for Spiritual Leadership." 1932. *PPA* 4: 2039–44.

————. "Some Responsibilities of Science with Relation to Government." 1934. *PPA* 4: 2480–89.

————. "Spiritual Values and the Constructive Life." 1933. *PPA* 4: 2051–60.

————. "Ultimate Values of Science." 1935. *PPA* 4: 2104–15.

Millikan, Robert Andrews. "Albert A. Michelson." *Science*, n.s. 73 (May 22, 1931): 549–50.

————. "Alleged Sins of Science." *Scribner's Magazine* 87 (February 1930): 118–29. Also in *Review of Reviews* 81 (March 1930): 94.

————. "Atomic Theory of Electricity." *Independent* 72 (June 13, 1912): 1302–08.

————. "Autobiographical Note on Work." *Science*, n.s. 73 (April 10, 1931): 386.

————. "Available Energy." *Science*, n.s. 68 (September 28, 1928): 279–84.

————. "The Birth of Two Ideas: What They Do for Us Today." *Scribner's Magazine* 80 (November 1926): 555–59.

————. "Contributions of Physical Science." In *The New World of Science: Its Development During the War*, edited by Robert M. Yerkes. New York: Century, 1920.

————. "Contributions to a British Association Discussion on the Evolution of the Universe." *Nature* 128 (October 24, 1931): 709–15.

————. "Correlation of High School and College Physics." *School Science and Mathematics* 9 (May 1909): 466–73.

————. "The Diffusion of Science: The Natural Sciences." *Scientific Monthly* 35 (September 1932): 203–08.

————. "Edison as a Scientist." *Science*, n.s. 75 (January 15, 1932): 68–70.

————. "Education and Unemployment." *Atlantic Monthly* 148 (December 1931): 804–10.

————. "Elimination of Waste in the Teaching of High School Science." *School and Science* 3 (January 29, 1916): 162–69. Also in *School Science and Mathematics* 16 (March 1916): 193–202.

————. "Evolution of Twentieth Century Physics." *Annual Report of the Board of Regents of the Smithsonian Institution*. Washington, D.C.: Government Printing Office, 1927, pp. 191–99.

————. "Growth." *American Magazine* 119 (January 1935): 7.

————. "Gulliver's Travels in Science." *Scribner's Magazine* 74 (November 1923): 577–85.

———. "High Frequency Rays of Cosmic Origin." *Science*, n.s. 62 (November 20, 1925): 445–48. Also in *Nature* 116 (December 5, 1925): 823–25; *Scientific Monthly* 21 (December 1925): 661–64; *Proceedings of the National Academy of Sciences* 12 (January 1926): 48–55; *Popular Astronomy* 34 (April 1926): 232–38; *Annual Report of the Board of Regents of the Smithsonian Institution.* Washington, D.C.: Government Printing Office, 1926, pp. 193–201.

———. "I Believe in God." *Scribner's Commentator* 11 (January 1942): 6–8.

———. "The Last Fifteen Years of Physics." *Proceedings of the American Philosophical Society* 65, no. 2 (1926): 68–78.

———. "Letter." *Outlook* 159 (September 2, 1931): 9.

———. "Michelson's Economic Value." *Science*, n.s. 69 (May 10, 1929): 481–85.

———. "Modern Alchemy." *Atlantic Monthly* 157 (June 1936): 737–41.

———. "Modern Physics." In *Contemporary Science*, edited by Benjamin Harrow. New York: Boni and Liveright, 1921.

———. "New Frontiers of Economic Progress: The Fallacies of Marx." *Vital Speeches of the Day* 4 (June 15, 1938): 537–40.

———. "The New Opportunities in Science." *Science*, n.s. 50 (September 26, 1919): 285–97.

———. "The New Physics." *School Review* 23 (November 1915): 607–20.

———. [Philosophy]. In *Living Philosophies*, edited by Albert Einstein et al. New York: Simon and Schuster, 1931.

———. "The Practical Value of Pure Science." *Science*, n.s. 59 (January 4, 1924): 7–10.

———. "The Present Needs of Science Instruction in Secondary Schools." *School Science and Mathematics* 20 (February 1920): 101–04.

———. "Present Status of Theory and Experiment as to Atomic Disintegration and Atomic Synthesis." *Science*, n.s. 73 (January 2, 1931): 1–5. Also in *Nature* 127 (January 31, 1931): 167–70; and *Annual Report of the Board of Regents of the Smithsonian Institution.* Washington, D.C.: Government Printing Office, 1931, pp. 277–85.

———. "The Problem of Science Teaching in the Secondary Schools." *School and Society* 22 (November 21, 1925): 633–39. Also in *School Science and Mathematics* 25 (December 1925): 966–75.

———. "Recent Discoveries in Radiation and Their Significance." *Popular Science Monthly* 64 (April 1904): 481–99.

———. "Relations of Science to Industry." *Science*, n.s. 69 (January 11, 1929): 27–31.

———. "Science and Human Affairs—Abstract." *Addresses and Proceedings, National Education Association* 61 (1923): 845–48.

———. "Science and Modern Life." *Atlantic Monthly* 141 (April 1928): 487–96. Also in *The Creative Intelligence and Modern Life*, by Francis John McConnell, Frederick J. E. Woodbridge, Roscoe Pound, Lorado Taft, Robert A. Millikan, and Paul E. Shorey. Boulder: University of Colorado Press, 1928.

———. "Science and Social Justice: 'A Stupendous Amount of Woefully Crooked Thinking.'" *Vital Speeches of the Day* 5 (December 1, 1938): 98–101.

———. "Science and Society." *Science*, n.s. 58 (October 19, 1923): 293–98.

———. "Science and the Standard of Living." *Forum* 99 (March 1938): 171–76.

———. "Science and the World of Tomorrow." *Vital Speeches of the Day* 5 (May 1, 1939): 446–48. Also in *Scientific Monthly* 49 (September 1939): 210–14, and *Reference Shelf* 13, no. 3 (1939): 187–96.

————. "Science in the Secondary Schools." *School Science and Mathematics* 17 (May 1917): 379–87.

————. "A Scientist Confesses His Faith." *Christian Century* 40 (June 21, 1923): 778–83. Also issued as a pamphlet: *A Scientist Confesses His Faith.*

————. "Seeing the Invisible." *Scribner's Magazine* 74 (October 1923): 445–57.

————. "The Significance of Radium." *Science*, n.s. 54 (July 1, 1921): 1–8.

————. "Some Scientific Aspects of the Meteorological Work of the United States Army." *Proceedings of the American Philosophical Society* 58 (1919): 133–49. Also in *The New World of Science: Its Development During the War*, edited by Robert M. Yerkes. New York: Century, 1920.

————. "Twentieth Century Physics." *Annual Report of the Board of Regents of the Smithsonian Institution.* Washington, D.C.: Government Printing Office, 1918, pp. 169–84.

————. "What I Believe." *Forum* 82 (October 1929): 193–201.

Millikan, Robert Andrews, and Cameron, G. Harvey. "Direct Evidence of Atom Building." *Science*, n.s. 67 (April 13, 1928): 401–02.

————. "Evidence for the Continuous Creation of the Common Elements out of Positive and Negative Electrons." *Proceedings of the National Academy of Sciences* 14 (June 1928): 445–50.

————. "Evidence That the Cosmic Rays Originate in Interstellar Space." *Proceedings of the National Academy of Sciences* 14 (August 1928): 637–41.

M[ore]., L[ouis]. T[renchard]. "The Theory of Relativity." *Nation* 94 (April 11, 1912): 370–71.

Münsterberg, Hugo. "The St. Louis Congress of Arts and Sciences." *Atlantic Monthly* 91 (May 1903): 671–84.

"A New and Revolutionary Doctrine of Time and Space." *Current Opinion* 63 (September 1917): 178.

"Popular Appreciation of Scientists." *Nation* 74 (January 16, 1902): 46–47.

"Popular Science." *Scientific American* 105 (October 21, 1911): 362.

Poor, Charles Lane. "Is Einstein Wrong?—A Debate: I. The Errors of Einstein." *Forum* 71 (June 1924): 705–15.

————. "Planetary Motions and the Einstein Theories: A Possible Alternative to the Doctrines That Would Save the Newtonian Law." *Scientific American Monthly* 3 (June 1921): 484–86.

"Repudiation of Common Sense by the New Physics: Has the Paradox of Matter and Motion Been Carried Too Far?" *Current Opinion* 64 (June 1918): 406–07.

Ritter, William Emerson. "The Relation of E. W. Scripps to Science." *Science*, n.s. 65 (March 25, 1927): 291–92.

Robertson, T. Brailsford. "The Cash Value of Scientific Research." *Scientific Monthly* 1 (November 1915): 140–47.

"Safety on the Railroads." *Century* 74 (June 1907): 321.

"Scientists and the Masses." *Nation* 92 (May 4, 1911): 441–42.

Skidmore, Sidney T. "The Mistakes of Dr. Einstein." *Forum* 66 (August 1921): 119–31.

Slosson, Edwin E. "Action and Reaction in Spreading Science." *School and Society* 23 (February 20, 1926): 223–30.

————. "Back to Nature? Never! Forward to the Machine." *Independent* 101 (January 3, 1920): 5.

————. "The Best Is Yet to Be." *Collier's Weekly* 80 (October 22, 1927): 8–9.

————. "Can You Tell the Difference Between Rest and Motion? Does the Earth Move Round the Sun or the Sun Move Round the Earth? Do Two Parallel Lines Ever Meet? Do We Need a Fourth Dimension?" *Independent* 100 (December 13, 1919): 174–75.

————. "Chemical and Industrial Mobilization." *Proceedings of the Academy of Political Science* 12 (July 1926): 73–78.

————. "Chemistry and the Past, with Interesting Predictions as to the Future." *Century* 120 (January 1930): 29–37.

————. "Chemistry in Everyday Life." *Mentor* (Springfield, Ohio) 10 (April 1922): 2–12.

————. "The Chemistry of the Greatest Miracle in the Bible." *Independent* 55 (June 18, 1903): 1454–57.

————. "A Clearing House of the Sciences." *Independent* 57 (October 6, 1904): 788–94.

————. "Coming of the New Coal Age." *Annual Report of the Board of Regents of the Smithsonian Institution.* Washington, D.C.: Government Printing Office, 1927, pp. 243–53.

————. "Creative Chemistry." *Independent* 92 (October 13, 1917): 91–93; (October 22, 1917): 185–86; (November 10, 1919): 291–92; (November 24, 1917): 378–81; (December 8, 1917): 476–77; (December 22, 1917): 556. *Independent* 93 (January 12, 1918): 70–71; (February 9, 1918): 237–38; (March 9, 1918): 416–17.

————. "Do the Papers Lie About Science?" *Scientific Monthly* 16 (May 1923): 559–60.

————. "Einstein's Crease." *Independent* 105 (January 8, 1921): 42–43.

————. "Einstein's Reception." *Independent* 105 (April 16, 1921): 400–01.

————. "The Ethics of Evolution." *World Today* 47 (February 1926): 196–99.

————. "Gasolene as a World Power." In *Science Remaking the World*, edited by Otis William Caldwell and E. E. Slosson. Garden City, N.Y.: Garden City Publishing Co., 1923.

————. "H. G. Wells, Social Prophet." *Independent* 76 (November 20, 1913): 348–53.

————. "How New Metals Are Made." *Independent* 101 (January 24, 1920): 136.

————. "If We Knew as Much as a Tree: The Big Problem of Applied Science—How to Put the Sun to Work." *Independent* 107 (October 1, 1921): 14.

[————.] "The Impending Subjugation of Nature." *Current Opinion* 70 (March 1921): 369–70.

————. "The Influence of Coal-Tar on Civilization." In *Science Remaking the World*, edited by Otis William Caldwell and E. E. Slosson. Garden City, N.Y.: Garden City Publishing Co., 1923.

————. "Inventory of Energy." *Scientific Monthly* 16 (February 1923): 217–19.

————. "Is There a Law of Human Progress? Speculations on the Acceleration of Scientific Knowledge." *Independent* 108 (February 25, 1922): 185–86.

————. "John Dewey: Teacher of Teachers." *Independent* 89 (March 26, 1917): 541–44.

————. "Live History." *Independent* 87 (September 11, 1916): 384.

————. "Madame Curie and Her Gram of Radium." *Independent* 105 (June 4, 1921): 584–85.

————. "Michelson's Eye." *Independent* 105 (January 15, 1921): 64–65.

————. "Motion: A Reverie Instigated by the Restlessness of the Waves." *Independent* 83 (August 2, 1915): 154–56.

————. "The Narrowing of Words and the Widening of Minds." *Columbia University Teachers College Record* 24 (May 1923): 197–203.

————. "A New Agency for the Popularization of Science." *Science*, n.s. 53 (April 8, 1921): 321–23.

————. "New Wonders of Science." *Independent* 108 (May 13, 1922): 444–46.

————. "A Number of Things." *Independent* 84 (November 15, 1915): 288.

————. "On Translating Einstein." *Science*, n.s. 56 (December 29, 1922): 752–54.

————. "The Philosophy of General Science." *School and Society* 20 (December 27, 1924): 799–806.

————. "Pragmatism." *Independent* 62 (February 21, 1907): 422–25.

————. "Research plus News-Gathering." *Science*, n.s. 57 (June 22, 1928): 628–29.

————. "Science and Journalism: The Opportunity and the Need for Writers of Popular Science." *Independent* 74 (April 24, 1913): 913–18.

————. "Science for All Sorts and Conditions of Men." *Bookman* 63 (March 1926): 67–73.

————. "Science for the Million." *Addresses and Proceedings, National Education Association* 62 (1924): 754–61.

————. "Science from the Side-lines." *Century* 103 (January 1922): 471–76.

————. "Science Remaking the World." *World's Work* 45: "I. Wonder-Working Gasolene" (November 1922): 39–50; "II. Cold Almost as Useful as Heat" (December 1922): 162–75; "III. Coal Tar as a World Power" (January 1923): 255–65; "IV. The Influence of Photography on Modern Life" (February 1923): 399–416; "V. The Influence of Sugar-Power in History" (March 1923): 495–508.

————. "Science Teaching in a Democracy." *School and Society* 19 (April 5, 1924): 383–88.

————. "Science Vs. Literature as a Professorial Profession." *Independent* 69 (December 29, 1910): 1440–42. Also in *Cap and Gown* (University of Chicago), 1910.

————. "Scientific Obscurity." *Science*, n.s. 67 (April 13, 1928): 398.

————. "Sun Dogs." *Independent* 104 (October 2, 1920): 10.

————. "Survival of the Unfittest." *Independent* 107 (October 8, 1921): 24–25.

————. "That Elusive Fourth Dimension." *Independent* 100 (December 27, 1919): 274.

————. "Things You Can't Be Sure Of." *Independent* 100 (December 20, 1919): 236.

————. "The Weight of Light." *Independent* 100 (November 29, 1919): 136.

————. "Wells on the World." *Independent* 104 (December 11, 1920): 361–62.

————. "What Science Means in Our Lives." Abridged. *World Review* 5 (October 24, 1927): 90–91.

————. "Will the Moon Explode?" *Collier's Weekly* 81 (April 7, 1928): 53.

————. "The Writing of Popular Science." *Science*, n.s. 55 (March 3, 1922): 241–42. A discussion of this article can be followed in *Science*, n.s. 55 (April 7, April 28, May 5, June 2, June 9, 1922): 374–76, 454–55, 480–82, 593–95, 620–21.

Stewart, John Q. "The Nature of Things: Einstein's Theory of Relativity—A Brief Statement of What It Is and What It Is Not." *Scientific American* 122 (January 3, 1920): 10.

Stowe, Lyman Beecher. "Patriots in the Public Service." *Outlook* 92 (July 24, 1909): 717–25.

Stratton, George Malcolm. "Railway Accidents and the Color Sense." *Popular Science Monthly* 72 (March 1908): 244–52.

————. "Railway Disasters at Night." *Century* 74 (May 1907): 118–23.

Townsend, E. J. "Science and Public Service." *Science*, n.s. 32 (November 4, 1910): 609–21.

W[a]lcott, C. D. "Mr. Carnegie's Gift to the Nation." *Independent* 53 (December 19, 1901): 2988–89. (The *Independent* incorrectly spelled the author's name as *Wolcott*.)

Webster, Arthur Gordon. "America's Intellectual Product." *Popular Science Monthly* 72 (March 1908): 193–210.

Wells, H. G. "The Discovery of the Future." *Annual Report of the Board of Regents of the Smithsonian Institution.* Washington, D.C.: Government Printing Office, 1903, pp. 375–92.

Secondary Published Sources

BOOKS

Armytage, W. H. G. *The Rise of the Technocrats; A Social History.* London: Routledge & Kegan Paul, 1965.

Bannister, Robert C., Jr. *Ray Stannard Baker: The Mind and Thought of a Progressive.* New Haven: Yale University Press, 1966.

Bates, Ralph Samuel. *Scientific Societies in the United States.* 3d ed. Cambridge, Mass.: M.I.T. Press, 1965.

Bohr, Niels. *Atomic Theory and the Description of Nature.* Cambridge, England: University Press, 1961.

Broderick, John Thomas. *Willis Rodney Whitney: Pioneer of Industrial Research.* Albany, N.Y.: Fort Orange Press, 1946.

Buckley, Jerome Hamilton. *The Triumph of Time: A Study of the Victorian Concepts of Time, History, Progress, and Decadence.* Cambridge, Mass.: Harvard University Press, Belknap Press, 1966.

Bury, J. B. *The Idea of Progress: An Inquiry into Its Origins and Growth.* Preface by Charles A. Beard. 1932. Reprint. New York: Dover Publications, 1955.

Calvert, Monte A. *The Mechanical Engineer in America, 1830–1910: Professional Cultures in Conflict.* Baltimore: Johns Hopkins Press, 1967.

Capek, Milič. *The Philosophical Impact of Contemporary Physics.* Princeton, N.J.: Van Nostrand, 1961.

Childs, Herbert. *An American Genius: The Life of Ernest Orlando Lawrence, Father of the Cyclotron.* New York: E. P. Dutton, 1968.

Commager, Henry Steele. *The American Mind: An Interpretation of American Thought and Character Since the 1880's.* New Haven: Yale University Press, 1950.

Copley, Frank Barkley. *Frederick W. Taylor, Father of Scientific Management.* 2 vols. New York: Harper & Brothers, 1923.

Corner, George W. *A History of the Rockefeller Institute, 1901–1953: Origins and Growth.* New York: Rockefeller Institute Press, 1964.

Coulson, Thomas. *Joseph Henry: His Life and Work.* Princeton: Princeton University Press, 1950.

Cremin, Lawrence A. *The Transformation of the School: Progressivism in American Education, 1876–1957.* New York: Alfred A. Knopf, 1961.

Danielian, N. R. *A.T.&T.: The Story of Industrial Conquest.* New York: Vanguard Press, 1939.

Daniels, George H. *American Science in the Age of Jackson*. New York: Columbia University Press, 1968.

Davis, Nuel Pharr. *Lawrence and Oppenheimer*. New York: Simon and Schuster, 1968.

Dupree, A. Hunter. *Science in the Federal Government: A History of Policies and Activities to 1940*. New York: Harper & Row, Torchbooks, 1964.

Dutton, William S. *Du Pont: One Hundred and Forty Years*. New York: Charles Scribner's Sons, 1942.

Ellul, Jacques. *The Technological Society*. Translated by John Wilkinson. First edition in France 1954; in the United States 1964. Reprint. New York: Random House, Vintage Books, 1967.

Forcey, Charles. *The Crossroads of Liberalism: Croly, Weyl, Lippmann, and the Progressive Era, 1900–1925*. New York: Oxford University Press, 1967.

Gardner, Gilson. *Lusty Scripps: The Life of E. W. Scripps*. New York: Vanguard Press, 1932.

Gay, Peter. *The Enlightenment—An Interpretation: The Rise of Modern Paganism*. New York: Alfred A. Knopf, 1966.

Goldman, Eric. *Rendezvous with Destiny: A History of Modern American Reform*. New York: Alfred A. Knopf, 1952.

Gray, George W. *Science at War*. New York: Harper & Brothers, 1943.

Greenberg, Daniel S. *The Politics of Pure Science*. New York: New American Library, 1967.

Haber, Samuel. *Efficiency and Uplift: Scientific Management in the Progressive Era, 1890–1920*. Chicago: University of Chicago Press, 1964.

Hall, Courtney Robert. *History of American Industrial Science*. New York: Library Publishers, 1954.

Hayek, F. A. *The Counter-Revolution of Science: Studies on the Abuse of Reason*. 1955. Reprint. London: Free Press, Collier-Macmillan, 1964.

Hays, Samuel P. *Conservation and the Gospel of Efficiency: The Progressive Conservation Movement, 1890–1920*. Cambridge, Mass.: Harvard University Press, 1959.

Heisenberg, Werner. *Physics and Philosophy: The Revolution in Modern Science*. New York: Harper & Row, Harper Torchbooks, 1958.

Hindle, Brook. *The Pursuit of Science in Revolutionary America, 1735–1789*. Chapel Hill: Published for the Early American Institute of History and Culture, Williamsburg, Virginia, by the University of North Carolina Press, 1956.

Hofstadter, Richard. *Social Darwinism in American Thought*. Boston: Beacon Press, 1955.

Holland, Maurice, with Henry F. Pringle. *Industrial Explorers*. New York: Harper & Brothers, 1928.

Hook, Sidney, ed. *John Dewey: Philosopher of Science and Freedom*. New York: Dial, 1950.

Hopkins, N. M. *The Outlook for Research and Invention*. New York: Van Nostrand, 1919.

Hoyle, Fred. *Galaxies, Nuclei and Quasars*. New York: Harper & Row, 1965.

Jessup, Philip C. *Elihu Root*. 2 vols. New York: Dodd, Mead, 1938.

Johnson, Thomas Cary. *Scientific Interests in the Old South*. New York: D. Appleton Century, 1936.

Jones, Richard Foster. *Ancients and Moderns: A Study of the Rise of the Scientific Movement in Seventeenth-Century England*. 1936. Reprint. Berkeley: University of California Press, 1965.

Jordy, William H. *Henry Adams: Scientific Historian*. New Haven: Yale University Press, 1952.

Karl, Barry Dean. *Executive Reorganization and Reform in the New Deal: The Genesis of Administrative Management, 1900–1939*. Cambridge, Mass.: Harvard University Press, 1963.

Lemon, Harvey Brace. *Cosmic Rays Thus Far*. Foreword by Arthur Holly Compton. New York: W. W. Norton, 1936.

Lichtheim, George. *The Concept of Ideology and Other Essays*. New York: Random House, Vintage Books, 1967.

Lurie, Edward. *Louis Agassiz: A Life in Science*. 1960. Abridged edition. Chicago: University of Chicago Press, Phoenix Books, 1966.

Manning, Thomas G. *Government in Science: The U.S. Geological Survey, 1867–1894*. Lexington: University of Kentucky, 1967.

May, Henry F. *The End of American Innocence: A Study of the First Years of Our Own Time, 1912–1917*. Chicago: Quadrangle Paperbacks, 1964.

Miller, John Anderson. *Workshop of Engineers: The Story of the General Engineering Laboratory of the General Electric Company, 1895–1952*. Schenectady, N.Y.: General Electric Company, 1953.

Nadworny, Milton J. *Scientific Management and the Unions, 1900–1932*. Cambridge, Mass.: Harvard University Press, 1952.

Nieburg, H. C. *In the Name of Science*. Chicago: Quadrangle Books, 1966.

Noble, David. *The Paradox of Progressive Thought*. Minneapolis: University of Minnesota Press, 1958.

Schlipp, Paul Arthur, ed. *Albert Einstein: Philosopher-Scientist*. 1949. 2 vols. Reprint. New York: Harper & Row, 1959.

Sussman, Herbert L. *Victorians and the Machine: The Literary Response to Technology*. Cambridge, Mass.: Harvard University Press, 1968.

Sypher, Wylie. *Literature and Technology: The Alien Vision*. New York: Random House, 1968.

Toulmin, Stephen, and Goodfield, June. *The Discovery of Time*. 1965. Reprint. New York: Harper & Row, Harper Torchbooks, 1966.

Trombley, Kenneth E. *The Life and Times of a Happy Liberal: A Biography of Morris Llewellyn Cooke*. New York: Harper & Brothers, 1954.

True, Frederick W., ed. *A History of the First Half-Century of the National Academy of Sciences, 1863–1913*. Washington: National Academy of Sciences, 1913.

Visher, Stephen Sargent. *Scientists Starred, 1903–1943, in "American Men of Science": A Study of Collegiate and Doctoral Training, Birthplace, Distribution, Backgrounds, and Developmental Influences*. Baltimore: Johns Hopkins Press, 1947.

Weidlein, Edward R., and Hamor, William A. *Glances at Industrial Research During Walks and Talks in Mellon Institute*. New York: Reinhold Publishing, 1936.

Welter, Rush. *Popular Education and Democratic Thought in America.* New York: Columbia University Press, 1962.

White, Morton. *Social Thought in America: The Revolt Against Formalism*. Boston: Beacon Press, 1957.

Wiebe, Robert H. *The Search for Order: 1877–1920*. New York: Hill and Wang, 1967.

Wilson, Edwin Bidwell. *History of the Proceedings of the National Academy of Sciences, 1914–1963*. Washington, D.C.: National Academy of Sciences, 1966.

Wright, Helen. *Explorer of the Universe: A Biography of George Ellery Hale*. New York: E. P. Dutton, 1966.

ARTICLES

Auerbach, Lewis E. "Science in the New Deal: A Pre-War Episode in the Relations Between Science and Government in the United States." *Minerva* 3 (Summer 1965): 457–82.

Čapek, Milič. "The Theory of Eternal Recurrence in Modern Philosophy of Science, with Special Reference to C. S. Peirce." *Journal of Philosophy* 57 (April 28, 1960): 289–96.

Cohen, I. Bernard. "American Physicists at War: From the First World War to 1942." *American Journal of Physics* 13 (October 1945): 333–46.

———. "American Physicists at War: From the Revolution to the World Wars." *American Journal of Physics* 13 (August 1945): 223–35.

Greenberg, Daniel S. "The Politics of Pure Science." *Saturday Review* 50 (November 4, 1967): 62–79.

Holton, Gerald. "Influences on Einstein's Early Work in Relativity Theory." *American Scholar* 37 (Winter 1967–68): 59–79.

———. "Modern Science and the Intellectual Tradition." In *The New Scientist: Essays on the Methods and Values of Modern Science*, edited by Paul C. Obler and Herman A. Estrin. Garden City, N.Y.: Doubleday, Anchor Books, 1962.

Kaplan, Sidney. "Social Engineers as Saviors: Effects of World War I on Some American Liberals." *Journal of the History of Ideas* 17 (June 1956): 347–69.

Kennedy, Gail. "Science and the Transformation of Common Sense: The Basic Problem of Dewey's Philosophy." *Journal of Philosophy* 51 (May 27, 1954): 313–25.

Lear, John. "Science vs. Democracy: The Developing Struggle." *Saturday Review* 50 (November 4, 1967): 57–61.

Mann, Arthur. "The Progressive Tradition." In *The Reconstruction of American History*, edited by John Higham. New York: Harper & Row, Harper Torchbooks, 1962.

Mowry, George E. "Social Democracy, 1900–1918." In *The Comparative Approach to American History*, edited by C. Vann Woodward. New York: Basic Books, 1968.

Strout, Cushing. "The Twentieth Century Enlightenment." *American Political Science Review* 49 (1955): 321–39.

Wolfe, Dael. "Science and the Public Understanding." In *The New Scientist: Essays on the Methods and Values of Modern Science*, edited by Paul C. Obler and Herman A. Estrin. Garden City, N.Y.: Doubleday, Anchor Books, 1962.

Unpublished Secondary Sources

Layton, Edwin Thomas. "The American Engineering Profession and the Idea of Social Responsibility." Ph.D. dissertation, University of California, Los Angeles, 1956.

Kevles, Daniel Jerome. "The Study of Physics in America, 1865–1916." Ph.D. dissertation, Princeton University, 1964.

Swenson, Lloyd Sylvan. "The Ethereal Aether: A Descriptive History of the Michelson-Morley Aether-Drift Experiments, 1880–1930." Ph.D. dissertation, Claremont Graduate School, 1962.

INDEX

Abott, Edwin A., 118–22
Absolute simultaneity: and relativity theories, 130
Adams, Henry: and progress, 82, 136; and radium, 135
Agassiz, Louis: and popularization of science, 3; and evolution, 10; and National Academy of Sciences, 21; and scientific method, 163n48
Alfred College, 113
American Association for the Advancement of Science: in promoting research, 39; Committee on Policy of, 63; popular science journal of, 63, 64–66; and Science Service, 68–70; annual meeting of, 100
American Association of State Highway Officials, 60
American Chemical Society: popular science journal of, 63
American Iron and Steel Institute, 215
American Museum of Natural History, 27
American Physical Society, 100, 101
American Society for the Dissemination of Science: founded by E. W. Scripps, 66, 68
American Telephone and Telegraph: and National Research Fund, 207, 211, 214; patent suit for, 215
Angell, J. R., 69
Anthropology, American school of, 226
Antirationalism, 74
Anti-Saloon League, 72
Arabic, 35

Army Coast Defense, 43
Aston, Francis W., 141
Atomic bomb: Soddy prophesies, 149; Millikan denies possibility of, 151; Millikan's view of, 192–93; German, 229; and isolation of scientists, 230. *See also* Millikan, Robert Andrews; Soddy, Frederick
Atomic synthesis: Millikan's search for, 138–42

Bache, Alexander Dallas, 21
Bachelard, Gaston, 124
Bacon, Francis: philosophy of science of, 94; scientific method of, 157, 162; on science and authority, 226
Baekeland, L. H., 72
Baird, J. W.: and popular science, 8–9
Baker, Ray Stannard: and science, 94–95
Bannister, Robert, 94
Barrows, Albert L.: and impact of World War I on science, 59, 60; and peacetime national science, 60–61
Beard, Charles A.: as scientific progressive, 76; and scientific revolution, 86; and Millikan, 228
Becquerel, Antoine H., 134
Behaviorism: Watsonian, 75; Pavlovian, 226
Bell Telephone Laboratories, 214
Bergson, Henri: and relativity theory, 118n29; mentioned, 72
Bethe, Hans, 146
Biology, 31